Ruddy Ducks and Other Stifftails

Animal Natural History Series
Victor Hutchison, General Editor

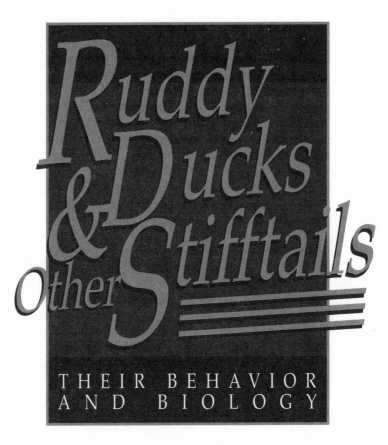

Ruddy Ducks & Other Stifftails

THEIR BEHAVIOR AND BIOLOGY

By
Paul A. Johnsgard
and
Montserrat Carbonell

University of Oklahoma Press : Norman and London

Library of Congress Cataloging-in-Publication Data

Johnsgard, Paul A.
    Ruddy ducks and other stifftails : their behavior and biology /
by Paul A. Johnsgard and Montserrat Carbonell.
        p.   cm. — (Animal natural history series : v. 1)
    Includes bibliographical references and index.
    ISBN 0–8061–2799–6 (alk. paper)
    1. Oxyura. 2. Anatidae.   I. Carbonell, Montserrat.   II. Title.
III. Series.
    QL696.A52J628   1996
    598.4'1—dc20   95–42400
                                                                      CIP

Ruddy Ducks and Other Stifftails: Their Behavior and Biology is Volume
1 in the *Animal Natural History Series.*

The paper in this book meets the guidelines for permanence and
durability of the Committee on Production Guidelines for Book
Longevity of the Council on Library Resources, Inc.⊗

        1   2   3   4   5   6   7   8   9   10

Contents

Illustrations

FIGURES

COLOR PLATES

Preface

By Paul A. Johnsgard

I can well remember the very first time that I saw a ruddy duck. I was in high school, and it was late April, which is still early spring in North Dakota. I was spending every spare day about 30 miles south of my home, watching and photographing waterfowl on a prairie marsh near Lake Traverse. It seemed to me that nearly all the waterfowl in central North America must gather there every spring on their annual migrations northward. On that particular day I had seen all the usual geese and ducks and was already overflowing with wonderful memories. Then, suddenly, I saw a small chestnut-brown duck swimming on the water, a bird with a long, cocked tail, a black head and immaculate white cheeks, and a sky-blue bill! Of course I knew immediately that it had to be a ruddy duck, and I was smitten instantly with this wonderful little bird. Since then I have been lucky enough see all of the seven other living species of stiff-tailed ducks, most of them in the wild as well as in captivity. The sighting of every new species has been an epiphany for me, much like watching Salome dance her dance of the seven veils, for observing each new species has allowed me to understand all of the others slightly better. Now I look forward each year to showing my students and friends ruddy ducks for the first time, for it is like introducing them to an old and valued friend, who I know will entertain as well enlighten them.

When I was in England in May of 1982 during one of my periodic visits to the Wildfowl Trust in Gloucestershire, England, I first

met Montse Carbonell. I myself had spent two years studying waterfowl at the Trust a decade previously, and at the time of my visit
she was working on a comprehensive study of stifftail behavior for
her doctoral research at Cardiff University. I told her that I thought
it was a wonderful research topic, and she reciprocated my interest
by sending me a copy of her 1983 dissertation after it was completed.
I later lost contact with her when she began a series of habitat studies across Latin America, one of which led to an important survey of
ecologically significant wetlands throughout the New World tropics
(Scott and Carbonell 1986).

In the spring of 1992, in thinking about possibilities for new
book projects, I toyed with the idea of a book on the biology of stiff-
tailed ducks. I quickly realized that I couldn't write such a book
without drawing heavily on Montse's unpublished dissertation and
I wrote to ask her if I could freely use her material in return for grant-
ing her coauthorship of the planned book. When she readily agreed
to this arrangement, I proceeded with the writing forthwith. Unless
otherwise indicated, unreferenced comments describing her observations relate specifically to her dissertation and associated information on captive birds at the Wildfowl Trust.

In the individual species accounts of social behavior, I have considerably condensed some of Montse's original display descriptions
and tabular data. I have also redrawn virtually all of her original line
illustrations. These drawings are based on a frame analysis of Super
8 cine sequences, as is the durational information provided for individual displays. I have also added many new drawings of my own,
mostly based on my own still photographs, cine sequences, or, in the
case of anatomical features, preserved museum specimens.

The assistance of several people must be gratefully acknowl-
edged. Dr. Barry Hughes very helpfully provided me with a copy of
his unpublished dissertation on the behavior of feral ruddy ducks in
England and sent me copies of some relatively inaccessible literature
references. In addition, he carefully read this manuscript and offered
many helpful comments. David Sharp of the U.S. Fish & Wildlife Service assembled and provided at my request several decades of unpublished data on hunter harvests and population surveys of North
American ruddy ducks. Photographs of wild masked ducks were offered to me by Rod Hall, Dirk Hagemeyer, and Dr. Humberto Alvarez. Rod Hall additionally loaned his transparencies of displaying
masked ducks for use as a documentary basis for my drawings and
provided me with some informal field notes on that elusive species.
Nigel Hewston also offered useful personal observations on stifftails
at the Wildfowl Trust.

My 1959–61 postdoctoral work on waterfowl behavior at the
Wildfowl Trust was supported by fellowships from the National Science Foundation and the U.S. Public Health Service. My field observations on stifftails in Australia during 1964 were greatly facilitated

by the late Vincent Lowe and his son Tom, by Harry Wright of the Victoria Fisheries and Wild Life Department, and by Dr. D. F. Dorward and the late Prof. A. J. Marshall, both of Monash University. My Australian field studies (1964), as well as those in South America (1965) and the West Indies (1966), were all supported by grants from the National Science Foundation.

Neither my work nor Montse's could have been done, or even begun, without the facilities and birds uniquely available at the Wildfowl Trust. Working there provided me with the two most professionally valuable and personally rewarding years of my entire career, and I am forever in debt to the late Sir Peter Scott for his lifelong efforts in establishing and developing the Trust. The Trust still remains a magical place for me, and it has come to represent Mecca for all ornithologists and other biologists who wish to understand the biology of the waterfowl family at a comprehensive or "holistic" level.

In recent years the Wildfowl Trust (which in 1989 was renamed the Wildfowl and Wetlands Trust but in the text is referred to by its earlier name) has expanded its interests to include the conservation of wetland habitats as well as of all waterfowl species. This was done in order to emphasize that waterfowl and their natural environments are inextricably linked, and one cannot conserve waterfowl without first conserving their wetland habitats. Global habitat-conservation problems are especially evident with regard to the stiff-tailed ducks. Nearly all of these species depend for their breeding on permanent wetlands that often are located in regions having valuable agricultural economies, such as native prairies. Increasing drainage and other ecological alterations of these wetlands are now putting the world's stiff-tailed ducks increasingly at risk, and this fact alone makes a summary of their biologies and ecologies of special interest and value at this time.

PAUL A. JOHNSGARD

Lincoln, Nebraska

Preface

By Montserrat Carbonell

Almost 10 years after putting away my dissertation thesis, when I had already given up the idea of publishing any papers on the stifftails I studied, Paul Johnsgard suggested preparing a book on this group of fascinating ducks. I promptly agreed, not only because—finally—my data could now become available to others but also because I hoped it would contribute to wetland conservation.

I have now lived in the Neotropics for 10 years, always hoping one day I may be able to look at some of the many intriguing aspects of the biology of the New World Oxyurini species; while this hasn't yet come true, I have managed to keep in touch with most of the research. Sadly I have noticed that, in spite of being one of the most interesting groups of ducks, with the exception only of the obligate nest parasite—the black-headed duck—very little attention has been paid to the stifftails by researchers, especially in South America. This may have a tragic result in the near future since wetlands on this continent are disappearing fast or are being influenced by human activity in a way with which most bird populations cannot cope. Pesticide runoff, drainage for agricultural purposes, transformation into rice paddies, garbage dumping, and water sports are some of the threats affecting wetlands in the Neotropics and elsewhere nowadays. If the distributions, population status, and biologies of waterfowl are not studied, we shall be able to do little to save the wetlands they depend upon.

Reading this book you will notice how little we know about

most species of stiff-tailed ducks, with the North American ruddy duck being perhaps the only exception. We hope that in raising more questions than answers we will stimulate some badly needed research.

MONTSERRAT CARBONELL

Gland, Switzerland

PART ONE

Comparative Biology

Male musk duck chasing water beetle. (After photo in Frith 1967.)

CHAPTER ONE

Stifftail Evolution and Taxonomy

The stiff-tailed ducks have traditionally been a relatively undisputed and well-defined group of eight to ten species in the waterfowl family Anatidae. Most members of this tribe, the Oxyurini (meaning sharp-tailed), may be easily distinguished from all other waterfowl by their elongated and pointed tail feathers with stiffened shafts. Their bills usually have flattened tips and narrow and variably recurved nails. These ducks also typically have very large, webbed feet and short legs that are placed far back on the body for maximum diving efficiency. Their short wings make takeoff difficult and flight labored. They usually have short, thick necks, which are variably inflatable in males during sexual display. Pair bonds tend to be weak and may even be nonexistent in some species. Vocalizations are few in both sexes, but displaying males often produce loud splashing noises using their feet, bill, wings, or tail. Their unusually dense and grebelike plumages are mostly composed of buff and grayish-brown (females) or ruddy-brown and sometimes blackish feathers (males). Contrasting white markings are mostly lacking or are limited to the underparts and occasionally the cheeks or faces of breeding males, who also tend to have ruddy-brown body plumages and generally blackish heads and necks. A white wing-speculum pattern is present in one species. The bill color of breeding males is also often a distinctive bright sky-blue color. Females of several stifftail species tend to lay their relatively large eggs in in nests of other birds (usually other ducks) as nesting "parasites," and their newly hatched ducklings are perhaps the correspondingly largest and most precocially independent of all waterfowl young (Johnsgard 1979).

As the stifftail group has been more recently defined (Johnsgard 1978, 1979), a total of eight extant species are recognized, including six in the "typical" genus *Oxyura* and one each in the variably "aberrant" genera *Heteronetta* and *Biziura*. Two other species, each representing a separate ("monotypic") genus (*Stictonetta* and *Thalassornis*), have at various times been regarded as stifftails or at least presumed to be related to them, but they are not so recognized in this book. The stifftails are widely distributed in wetland habitats throughout much of the world, but they are especially characteristic of tropical and subtropical swamps and marshes. Four of the stifftail species occur in South America, two are endemic to Australia, one occurs widely in North America (and also extends to South America), and one each is endemic to Africa and Eurasia (Figs. 1 and 2).

HISTORY OF STIFFTAIL CLASSIFICATION

To understand the present-day biological classification of stifftails, and to help perceive their relationships to one another and to other waterfowl groups, it is necessary to review the history of their study by waterfowl anatomists and other taxonomists.

In what was the first comprehensive classification of the Anatidae, Eyton (1838) recognized the stifftails as a unique group. He separated them as a distinct subfamily (Erismaturinae) containing three genera, *Erismatura (Oxyura), Biziura,* and *Thalassornis.* This last genus had been erected by Eyton for the newly discovered white-backed duck, which he first described and which he regarded as a connecting link between *Biziura* and the sea-duck genus *Clangula.* Eyton also believed that the typical stifftails show a slight anatomical approach to the mergansers, although his nonevolutionary taxonomic perspective was highly prone to confusing convergent anatomical similarities with actual phyletic relationships. The only later generic addition to the stifftail group was *Nomonyx,* which Ridgway (1880) proposed for distinguishing the masked duck from the other stifftails, mainly on the basis of its less recurved bill nail as compared with the narrow and recurved nail condition found in all other *Oxyura* species.

During post-Darwinian history, the most widely used classification of waterfowl has been one that was originally devised by Salvadori (1895) for classifying the birds represented in the collection of the British Museum (Natural History). Salvadori's seminal waterfowl classification, which had primarily utilized morphological and anatomical features of the feet and bill, was essentially a housekeeping classification designed for identifying and organizing museum collections. It was scarcely different from other waterfowl classifications of its day, such as that of Sclater (1880), and made little effort to integrate taxonomy with the then-emerging evolutionary concepts. The

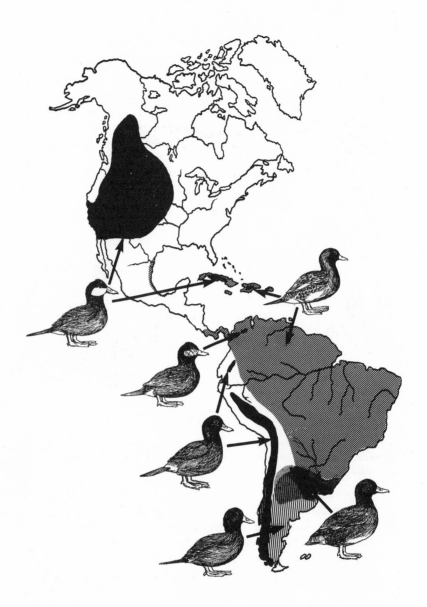

Figure 1. Collective breeding or residential distributions of the New World species of stifftails: Oxyura *species distribution is shown inked in or hatched;* Heteronetta *is shown stippled.*

stifftails were recognized by Salvadori as a distinct group containing four genera (*Thalassornis, Nomonyx, Erismatura,* and *Biziura*) and were placed sequentially between the subfamilies "Fuligulinae" (a group of diving-adapted ducks that included the pochards and the sea ducks other than the mergansers) and "Merganettinae" (the South American torrent ducks). Except for the recognition by Salvadori of

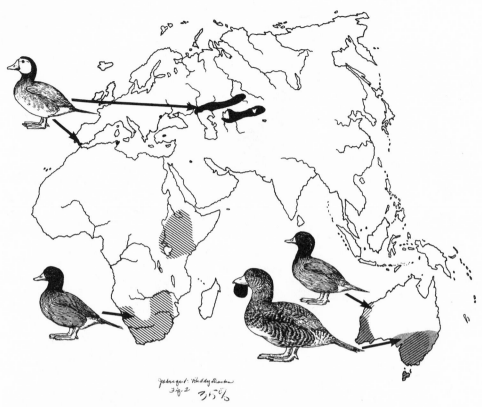

Figure 2. Collective breeding or residential distributions of the Old World species of stifftails: Oxyura *species distribution is shown inked in or hatched;* Biziura *is shown stippled.*

Nomonyx as a distinct genus, the composition of the group was not significantly modified from the classifications of Eyton and Sclater. It thus remained for Phillips (1922–26) to provide a more complete taxonomic synopsis of the stifftails and, more importantly, to provide the first comprehensive summary of what was then known of their natural histories.

In what was surely one of the most important monographs on any major bird group to be published during the twentieth century, Phillips's (1922–26) classic *A Natural History of the Ducks* generally followed Salvadori's classification, although the geese and swans were excluded from the book's coverage. One minor divergence from Salvadori's earlier waterfowl classification was that the whistling ducks were separated from the more typical ducks as a distinct subfamily (Dendrocygninae) by Phillips, who based this decision on anatomical research by Shufeldt (1914).

Following Salvadori's earlier classification, Phillips recognized the stiff-tailed ducks as composing a separate subfamily (Oxyurinae).

Within this subfamily a total of nine species was recognized, and these in turn were grouped into four genera (*Thalassornis, Nomonyx, Oxyura,* and *Biziura.*) Phillips collectively referred to this assemblage as the "spiny-tailed ducks." He characterized the group as having tail feathers with long, stiff shafts, extending well beyond the very short rump feathers; hind toes with large, flaplike lobes; and unusually thick necks, which in males often contain inflatable tracheal air sacs. In the first volume of his monumental taxonomy of the birds of the world only a few years later, Peters (1931) adopted the Salvadori-Phillips classification of Anatidae without significant alteration.

The first major advance toward a modern, evolutionarily based waterfowl classification finally occurred in 1945, when Jean Delacour and Ernst Mayr offered a radically new approach to waterfowl taxonomy. For the first time, a classification was proposed that utilized such features as behavioral traits, variations in natal and female plumages, and other presumably nonadaptive morphological characters. Using such traits helps to avoid problems of erroneously associating rather distantly related species that share common traits as a result of their having followed evolutionary pathways that convergently adapted them to similar ecologic habitats. Instead, using more fundamental or conservative traits produces better evidence of the species' real genetic affinities and of the relative degree of common ancestry. For example, some of the traits employed by Delacour and Mayr included adult vocalizations, mating systems and displays, and duckling plumage patterns. Anatomical traits that were used included the type of scaling pattern present on the surface of the tarsus—either reticulated (weblike) as in geese, swans, and whistling ducks or scutellated (vertically aligned) as in true ducks. Sexual and species differences in the structure of the trachea (windpipe) and syrinx (vocal organ) were also considered important clues to relationships by Delacour and Mayr. Major assemblages of related species were thus regarded as comprising seven tribes, which in turn were grouped into two large subfamilies, the Anserinae (geese, swans, and whistling ducks) and the Anatinae (sheldgeese, shelducks, and true ducks). In brief, Delacour and Mayr defined the subfamily Anatinae, and within that larger group the stiff-tailed duck tribe Oxyurini, as follows:

> Subfamily Anatinae: Tarsus scutellated anteriorly; a double annual molt; sexual dimorphism in adult plumages frequent; sexual dimorphism in voice and syringeal structure usual; mating displays typically elaborate and sexually different.
>
> Tribe Oxyurini (stiff-tailed ducks): Rectrices long and stiff, but tail coverts very short; bill broad and usually depressed, tipped with a sharp and recurved nail; legs placed far to the rear; neck short and very thick; wings (usually) lacking distinct speculum; no tracheal bulla; adults usually mute except

during sexual display; nests usually built of reeds; eggs very large; ducklings with an indistinct down pattern.

Whereas Phillips and earlier authors had sequentially placed the stifftails between the typical diving and sea ducks on the one hand and the mergansers on the other, Delacour and Mayr merged the mergansers with the sea ducks in a single tribe, Mergini, and placed the stifftails immediately after this combined assemblage. However, they diagrammatically depicted the stifftails as descending from a stem group of Anatinae that branched off early in anatine evolution. A comparison of the species-level taxonomy of stifftails proposed by Delacour and Mayr, relative to that initially established by Salvadori and later also used by Peters and Phillips, is of some historic interest. Such a comparison, using English vernacular names as employed by the authors cited, is seen in Table 1.

It may be seen that, in terms of the stifftails, Delacour and Mayr's classification was mainly innovative in that the black-headed duck was transferred into the stifftail group from its previous traditional position among the dabbling or surface-feeding ducks and was regarded as an "aberrant" member of the stifftail group. They also applied the term *aberrant* to the white-backed duck, noting several structural peculiarities, such as its very short tail feathers and its unusual downy plumage pattern. Additionally, Delacour and Mayr merged all of those species of *Oxyura* whose breeding males have entirely black head plumages into a single collective Southern Hemisphere species *australis*. This was done without any clear taxonomic justification (other than noting that the affected species are "similar in every respect") and without providing a suitable collective vernacular name for the entire group.

Delacour and Mayr's overall 1945 classification of the Anatidae was certainly a great improvement over all earlier ones and became essentially the basis for nearly all later variant classification schemes. So far as the stiff-tailed ducks are concerned, Delacour's (1959) later revision of this classification produced only minor taxonomic changes. It may be summarized as to its taxonomic sequence, species limits, and the Latin and English vernacular names utilized by Delacour as follows:

Oxyura (stiff-tailed ducks)
    O. jamaicensis: ruddy ducks, including North American,
    Andean (*vittata*) and Peruvian (*ferruginea*) races
    O. vittata: Argentine ruddy duck
    O. maccoa: maccoa duck
    O. australis: blue-billed duck
    O. leucocephala: white-headed duck
    O. dominica: masked duck
Thalassornis (white-backed ducks)

TABLE 1
Species-Level Taxonomy of Stifftails

Spiny-Tailed Ducks (Oxyurinae) after Phillips (1992–26)	Stiff-Tailed Ducks (Oxyurini) after Delacour and Mayr (1945)

Thalassornis
  T. leuconotus (white-backed duck)
	Oxyura
	  O. dominica (masked duck)
	  O. leucocephala (white-headed duck)
	  O. jamaicensis (North American ruddy duck)
	  O. australis (including *maccoa, ferruginea,* and *vittata*)

Nomonyx *N. dominicus* (masked duck)	

Oxyura
  O. leucocephala (white-headed duck)
  O. jamaicensis (ruddy duck)
  O. maccoa (maccoa duck)
  O. australis (blue-billed duck)
  O. ferruginea (Peruvian ruddy duck)
  O. vittata (Argentine ruddy duck)
	Biziura
	  B. lobata (Australian musk duck)

	Thalassornis *T. leuconota* (African white-backed duck)

Biziura
  B. lobata (musk duck)
	Heteronetta
	  H. atricapilla (black-headed duck)

T. leuconotus: African and Madagascan races of white-backed duck
Biziura (musk duck)
  B. lobata: Australian musk duck
Heteronetta (black-headed duck)
  H. atricapilla: black-headed duck

Differences in this listing from the earlier Delacour and Mayr classification mainly consist of a reversion of species limits in *Oxyura* to essentially that of the earlier traditional classifications, except that *ferruginea* was considered by Delacour to represent a subspecies of *jamaicensis* rather than a distinct species itself.

In 1961 Johnsgard suggested a number of changes in the classification of the Anatidae, based on his comparative behavioral observations of most species in the family. With regard to the stifftails, Johnsgard established the following points:

1. The black-headed duck is the least specialized species anatomically; behaviorally it links the stifftails to the typical surface-feeding or "dabbling" ducks of the tribe Anatini.
2. The white-backed duck only very doubtfully belongs within the stifftail tribe, and some structural and behavioral simiarities exist with the whistling ducks *(Dendrocygna).*
3. The six species of *Oxyura* recognized by Delacour in 1959 are well defined, but there is too little behavioral information to establish intrageneric affinities with any assurance.

In the same year, Woolfenden (1961) completed a study of the postcranial osteology of the waterfowl family. Although *Thalassornis* was unavailable for study, Woolfenden offered several points with regard to the remaining stifftails:

1. The genera *Heteronetta*, *Nomonyx*, *Oxyura*, and *Biziura* form a distinct tribe.
2. Of these genera, *Heteronetta* is least specialized and may form a link with the dabbling ducks.
3. The genus *Nomonyx* is similar to *Oxyura* but sufficiently distinctive to be recognized.
4. *Oxyura* and *Biziura*, and especially the latter, are the most specialized osteologically.

In 1965 Johnsgard published a taxonomy of the Anatidae based largely on his behavioral observations, which modified the Delacourian classification of stifftails only slightly, mainly by suggesting changes in linear sequence, from presumably more generalized to more specialized types:

Tribe Oxyurini (stiff-tailed ducks)

Heteronetta atricapilla (black-headed duck) *Oxyura*

Subgenus *Nomonyx*

O. (Nomonyx) dominica (masked duck)

Subgenus *Oxyura*

O. jamaicensis (ruddy duck)

O. leucocephala (white-headed duck)

O. maccoa (maccoa duck)

O. vittata (Argentine ruddy duck)

O. australis (Australian blue-billed duck)

Biziura lobata (musk duck)

Thalassornis leuconotus (white-backed duck)

The next development in the taxonomy and classification of the stifftails came about as a result of observations by Johnsgard (1967) on a breeding captive pair of white-backed ducks, together with a summary of his observations on the behavior of a variety of other stifftails. Johnsgard concluded that:

1. Several behavioral and anatomical lines of evidence indicate that the white-backed duck should be removed from the stifftail assemblage and regarded as an aberrant member of the whistling duck tribe (Dendrocygnini).
2. The black-headed duck is the most generalized of the stifftails, and it exhibits some pair-forming and pair-bonding characteristics that help in understanding the social behaviors of more typical stifftails.
3. The masked duck is probably the most isolated of the typical stifftails, but there is too little information on it to advocate generic distinction on that basis.

4. Remaining species in the genus *Oxyura* seem to fall into two broad evolutionary groups, including *leucocephala* and *jamaicensis* on the one hand and *maccoa, vittata,* and *australis* on the other.
5. The seemingly aberrant features of musk duck morphology and behavior can be attributed to the effects of intense sexual selection resulting from a promiscuous mating system in that species.

Of these conclusions, the splitting of the typical *Oxyura* species into two multispecies groups, including a somewhat more northern-oriented group (white-headed duck and ruddy duck) and a distinctly southern or austral one (the remaining three species), has since been reevaluated on the basis of new information. After observing the displays of the maccoa duck, Johnsgard (1968a) judged that the maccoa might be transitional toward the other southern forms rather than part of that group, but Siegfried and van der Merwe (1975) supported Johnsgard's original contention that the the austral species do represent a homogenous group. Carbonell (1983) similarly concluded that these austral species represent a single evolutionary cluster, which collectively are more closely related to the ruddy duck than to the white-headed duck. Special attention has also been given to the supposed relationship between the white-headed and ruddy ducks, whose superficial male plumage similarities are not matched by their social displays (Johnsgard 1978). This distinction was supported by Carbonell (1983), who believed the white-headed duck is behaviorally transitional between the *Oxyura* species and the musk duck genus *Biziura*. However, some of the apparent similarities may result from the effects of intense sexual selection in a polygynous-promiscuous mating system, as described above.

Of Johnsgard's other major conclusions, that regarding the relationships of the white-backed duck was certainly most controversial, inasmuch as its position among the stifftails had never previously been challenged. However, this position was strongly supported by the white-backed's anatomical traits, such as a syrinx much like that of the whistling ducks and a tarsal pattern that is reticulated as in geese and whistling ducks rather than scutellated as in true anatine ducks. Several behavioral features, such as its copulatory behavior, also helped link the white-backed to the whistling ducks. Additionally, Kear (1967) noted that the calls of downy young white-backed ducks are similar to those of whistling ducks, and a few years later Raikow (1971) supported this taxonomic repositioning after undertaking a study of the postcranial osteology of the white-backed duck. Carbonell's (1983) comparative study of the stifftails did not include the white-backed duck.

Weller (1967f, 1968a) reported that feather and plumage traits, anatomy, and behavior of the black-headed duck all place it closer to

Oxyura than to *Anas*, although it is evidently descended from an ancient lineage that may have dual affinities. Raikow (1970) had also
examined species representing the three generally accepted genera
of stifftails *(Biziura, Oxyura, Heteronetta)* as to their skeletal and muscular adaptations for diving; he concluded that the black-headed duck
is the least specialized and is structurally intermediate in many respects between the typical dabbling ducks *(Anas)* and the advanced
stifftails *(Oxyura* and *Biziura)*. Raikow then concluded that these latter two genera are the derived products of a long and partly common
history of diving forms originating from nondiving ancestral stock,
probably much like *Anas*. Of these two typical stifftails, *Biziura* is the
more highly specialized for aquatic locomotion, especially in terms
of its hind limb and caudal musculature, the structure and proportions of its pelvic girdle, its tail length, and its hind limb proportions.
Carbonell (1983) supported the view that, on the basis of its behavior, the black-headed duck represents an evolutionary link between
the Anatini and Oxyurini.

In an important contribution to waterfowl classification, Brush
(1976) examined the feather proteins electrophoretically. With regard
to the stifftails, he concluded that the protein patterns of all four of
the species of *Oxyura* (including the masked duck) studied were identical to one another and also to those of the musk duck, but that the
proteins of the black-headed duck electrophoretically resembled those
of the dabbling ducks. The feather proteins of the white-backed duck
most resembled those of the whistling ducks. The stifftails were illustrated diagrammatically as being a sister group to the sea ducks
(eiders and scoters) rather than as being directly diverged from the
dabbling ducks.

In 1979 Johnsgard fully revised Peters's taxonomy of the Anatidae and developed a classification that was based on a review of all
the pertinent systematic information available at that time. Inasmuch
as the tribal category had not been used by Peters, the stifftails were
assigned the rank of subfamily, as in Peters's earlier taxonomy. However, *Thalassornis* was removed from the stifftail group and transferred
to the whistling ducks. The sequence of the remaining stifftail species
and genera remained as Johnsgard has recommended in earlier publications (1965a, 1978).

In 1983 Bottjer performed a cladistic analysis of the family using
immunological techniques. His data were obtained from blood samples representing 75 anatid species, which included a single stifftail,
the North American ruddy duck. Neither the white-backed duck
nor the black-headed duck was included in his survey. His major
taxonomic conclusion concerning the stifftails *(Oxyura)* was that
they should be raised to subfamily level (Oxyurinae) and sequentially placed between the subfamilies Anserinae and Anatinae. Bottjer cladistically diagrammed *Oxyura* as having branched off very shortly
after the lineage producing the Anserinae (geese and swans) but be-

fore the branching points for the shelducks and typical ducks (Anatinae). He also suggested that the whistling ducks and stifftails share some common, albeit primitive, features and that the anomaly of the white-backed duck's uncertain taxonomic affinities between these taxonomically divergent groups might be resolved if the two groups are really much more closely related than generally has been believed.

In 1983 Patton and Avise also performed a cladistic analysis of 26 waterfowl taxa, using electrophoretic data from 82 electromorph frequencies among 17 to 19 genetic loci. In this case, the only stifftail in the sample (the North American ruddy duck) clustered nearest a few sea-duck genera (*Clangula, Melanitta*) but also branched off cladistically from the line giving rise to the geese and swans well before the dabbling ducks, providing some support for Bottjer's position on the group's potentially ancient origin. More recently, DNA-DNA hybridization data (Sibley and Alquist 1990) suggest that the stifftails may have diverged shortly after *Dendrocygna* and before the geese and swans, thus supporting the view that the stifftails are among the most isolated of all waterfowl groups. Sibley and Munroe (1990) thus listed the stifftails immediately after the whistling ducks (separated as a full family Dendrocygnidae), as the first subfamily of the Anatidae. Using DNA-DNA hybridization data, Madsen, McHugh, and de Kloet (1988) also reported an early, preanatinae divergence of *Oxyura;* they showed a dendrogram placing it between the whistling ducks and the freckled duck as to DNA differences.

The most important contribution to waterfowl taxonomy in recent years has been that of Livezey (1986). He performed a cladistic analysis of species representing all the Recent genera of Anatidae, using 120 morphological characters and thus providing a useful comparison with the results obtained through the largely behavioral approach by Johnsgard (1965a). With regard to the stifftails and presumptive stifftails, Livezey concluded that:

1. *Thalassornis* is most similar to *Dendrocygna* but they are not sister genera. Instead they compose a grade, with *Thalassornis,* best placed sequentially between the whistling ducks and the goose-swan assemblage as a monotypic subfamily.
2. *Heteronetta* is the sister group to the more typical stifftails and belongs within the stifftail tribe Oxyurini.
3. *Biziura* is more closely related to typical *Oxyura* species than to the masked duck, which forms a sister group to *Oxyura-Biziura.*
4. The resurrection of the genus *Nomonyx* is desirable for clarifying intratribal relationships in the stifftails.
5. The suggested sequence of genera within the stifftails is *Heteronetta, Nomonyx, Oxyura,* and *Biziura.*
6. The Oxyurini are listed last in tribal sequence. They are a sister group to the sea ducks rather than being allied to the dabbling ducks or to *Stictonetta* (which is considered to rep-

resent a transitional subfamily between the Anserinae and
the more typically ducklike subfamilies).

Thus, as of 1990, the overall tribal composition of the stifftails
was seemingly still somewhat unsettled, with the black-headed duck
peripherally placed in either the stifftail group (Johnsgard, Woolfenden, Weller, Raikow, Livezey) or more closely allied to the dabbling ducks (Brush). A few biologists (e.g., Rees and Hillgarth 1984)
have suggested a possible separate tribal allocation for this species,
which thus helps evade the question of its nearest affinities. Curiously,
Sibley and Munroe (1990) listed the black-headed duck as the final
species in their anatid sequence, following the mergansers and at
the opposite end of the family from the stifftails. The typical stifftails
have most often been regarded as being derived from other generalized anatine ducks, a view apparently shared by Delacour and Mayr
(1945). Others have since regarded the stifftails as probably having a
common ancestral origin with the sea ducks (as advocated by Brush
1976 and Livezey 1986), from an *Anas* like dabbling-duck ancestor not
very different from present-day *Heteronetta* (as per Johnsgard, Woolfenden, and Raikow), or as resulting from a very early divergence
from anserine and/or dendrocygnine ancestors and before the splitting of the anatine ducks into recognizable groups (Bottjer; Sibley
and Alquist; Madsen, McHugh, and de Kloet). Finally, evidence had
gradually accumulated for excluding the white-backed duck from the
stifftail assemblage (Johnsgard, Raikow, Livezey).

While the white-backed duck was thus being eliminated from
the stifftail group because of its increasingly apparent whistling-duck
affinities, the puzzling Australian freckled duck *(Stictonetta naevosa)*
was increasingly examined and discussed by avian taxonomists. Until
1965, this strange dull-colored duck had generally been associated
with the dabbling ducks, although it was widely regarded as representing an aberrant and primitive anatine type of doubtful affinities
(Phillips 1922–26; Delacour and Mayr 1945). Johnsgard (1961a) and
Woolfenden (1961) both questioned this treatment and suggested that
it might instead have anserine affinities. Their opinions were based
on the then-limited available anatomical and biological information
on the freckled duck, and both judged that it should perhaps be given
tribal status in the swan-goose subfamily Anserinae. Frith (1964) later
noted that its downy young are swanlike in plumage pattern and that
its trachea and syrinx are also anserine. Johnsgard (1965b) concluded
from observing some of the social behavior patterns of wild birds
that the freckled duck is an extremely generalized species with many
primitive and apparently anserine features and probably warrants
tribal separation within the Anserinae. However, he also pointed out
some similarities in its bill shape and structure to those of the stifftails, especially the black-headed duck. He suggested that the freckled duck may have descended from primitive anatid stock that even-

tually gave rise to the true ducks of the subfamily Anatinae, and particularly to the stifftails.

Brush (1976) similarly judged the freckled duck to be primitive and unique on the basis of its feather proteins, resembling the whistling ducks less than the swan-goose assemblage. Like Johnsgard, he also advocated placing it in a monotypic tribe within the Anserinae. Nevertheless, Livezey (1986) judged it to be the last cladistic branch in the grade of waterfowl having reticulated tarsi and associated it as a monotypic subfamily (Stictonettinae) with a sister group (Anserinae) that included all the swans and geese. The whistling ducks were also given subfamily rank and more distantly placed in this cladistic lineage, with *Thalassornis* providing a linking form between these birds and the Anserinae, also as a monotypic subfamily. In a DNA-DNA hybridization study, Madsen, McHugh, and de Kloet (1988) reported that the freckled duck probably is only distantly related to the other Anatidae and that its lineage must have diverged early from the main waterfowl lineage. Their dendrogram situated it between *Oxyura* and eight representative species of swans, geese, and true ducks.

In a recent, detailed study of the social behavior of the freckled duck, Fullagar, Davey, and Rushton (1990) suggested that some behavioral similarities exist between the displays and vocalizations of this species and such stifftails as the musk duck (call structure and rhythm similarities) and the black-headed duck (similarities with male's toad-call display). Preflight signals of the freckled duck are ducklike, not gooselike. Likewise, pair bonds of freckled ducks are short and seasonal, their vocalizations and ritualized displays are limited, and the displays do not include any mutual triumph ceremonies as had been earlier suspected (Johnsgard 1967). These and other behavioral traits suggested anatine rather than anserine affinities for the freckled duck to Fullagar Davey, and Rushton.

Using these findings and other recently available biological information, Marchant and Higgins (1990) have gone so far as to include the freckled duck within a redefined subfamily Oxyurinae. In part they characterized this reorganized subfamily (which also included *Biziura* and *Heteronetta* but excluded *Thalassornis*) as species having very short tail coverts and generally having a bill that is unusually widened and flattened at the tip, ending in a narrow, rather sharp nail. The young of freckled ducks and at least some of the *Oxyura* species also typically share a complete postjuvenile molt, a feature that is apparently unique to the group. In several species of stifftails, as well as in the freckled duck, adult males also uniquely have an extremely long "pseudopenis" that is greater in length than that of any other known bird. Males of most species (all but the musk duck) also have bright blue or basally reddish bills during breeding. Females are mostly silent, lacking both typical anatinae inciting and decrescendo calls.

It is perhaps too early to adopt this still-untested but intriguing proposition and to include without question the freckled duck among the stifftails. However, this possibility does add some weight to the view advanced by Bottjer (1983) that the stifftails may represent descendants of a very early evolutionary offshoot of the Anatidae, rather than being derived from more specialized dabbling-duck or diving-duck ancestral stock. In such a scenario the apparent connections of the black-headed duck with the typical dabbling ducks *(Anas)* must be regarded as representing chance similarities; alternatively, it might be claimed that the black-headed duck actually belongs with the dabbling ducks and thus has no close taxonomic affinities with the stifftails.

CURRENT STATUS

The major points of taxonomic contention and uncertainty still existing in the stifftails are as follows:

1. Is the white-backed duck a stifftail or a whistling duck? Nearly all recent studies favor the latter position or at least indicate that *Thalassornis* is not a stifftail.

2. How is the freckled duck related to the stifftails? If it is indeed a primitive stifftail or even the group's nearest relative, it may be necessary to relocate the stifftail tribe and insert it near the geese, swans, and whistling ducks. This was done by Sibley and Monroe (1990), who placed the stifftails between the whistling ducks and the freckled duck in taxonomic sequence. However, Livezey (1995) has questioned this conclusion, stating that the available evidence does not support this position, which was at least in part based on a misinterpretation by Fullagar (in Marchant and Higgins 1990) of a cladistic study by Faith (1989).

3. Is the black-headed duck more properly included with the stifftails or with the dabbling ducks? The general predominance of evidence is that it is closer to the stifftails, but it may be a transitional link with the Anatini—or possibly even a member of that group.

4. Does the masked duck warrant generic separation from the typical *Oxyura* stifftails? This is a quantitative rather than qualitative question and is not of great taxonomic significance. Too little information is available to decide this point, which can be managed by listing the species first in sequence among the typical stifftails and assigning it subgeneric ranking until more information becomes available.

5. What is the relationship of the musk duck to the smaller *Oxyura* stifftails? Although certainly a bizarre species, its secondary sexual attributes are understandable in the light of sexual selection influences, and it is clearly a very close but more anatomically specialized relative of *Oxyura*.

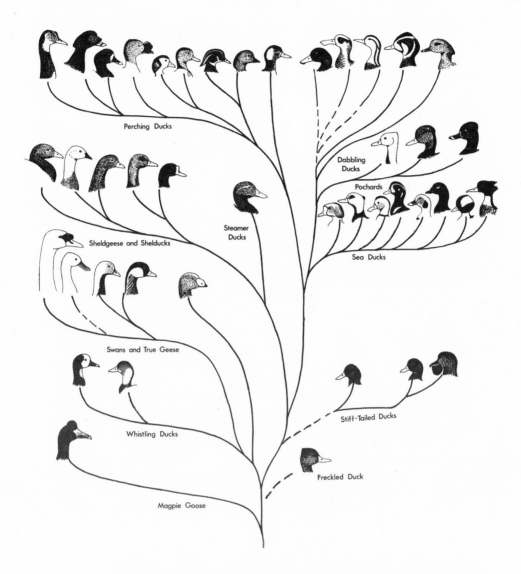

Figure 3. Generalized evolutionary dendrogram of the Anatidae, showing representative genera of all tribes and the approximate apparent phyletic position of the stiff-tailed ducks. (Modified from Johnsgard 1968b.)

In summary, this book adopts Johnsgard's (1978, 1979) taxonomic delineation of the stifftails as still being the most consistently in agreement with currently available anatomical, behavioral, and biochemical information and uses the same linear arrangement of species, except for a shift in the position of the white-headed duck. However, some comments on the behavior and biology of the white-backed duck and the freckled duck will be made wherever they seem to be relevant. Wherever the term *stifftails* or *stiff-tailed ducks* is used, it will be meant to encompass the genera *Heteronetta, Oxyura,* and *Biziura* collectively; similarly, references to the more restrictive "typical" stiff-tails will taxonomically encompass only *Oxyura* and *Biziura.* The term *ruddy ducks* is seemingly best limited by tradition to the *jamaicensis, andina, ferruginea* assemblage. The South American species *vitatta* has sometimes also been called the Argentine ruddy duck (Phillips 1922–26). This species has also sometimes been called the Argentine lake duck, but it is more closely associated with reedy marshes and ponds than with deep lakes. Since its taxonomic affinities are almost certainly much closer to those of the Australian blue-billed duck than to the ruddy ducks as defined above, it is here called the Argentine blue-billed duck.

More generally, in this book the term *sea ducks* refers collectively to the eiders, scoters, and mergansers of the tribe Mergini. *Pochards* likewise refers collectively to species of *Netta* and *Aythya* of the tribe Aythyini, a group that in North America is often called *bay ducks* or *inland divers.* However, *diving ducks* is used here as a general and broad descriptive term that encompasses the pochards, sea ducks, and all other diving-adapted waterfowl. The alternative terms *dabbling* or *surface-feeding* ducks encompass all the members of the tribe Anatini, especially those of the genus *Anas.* Vernacular English names used hereafter in the text for stifftails or other waterfowl species are those that were adopted by Johnsgard (1978). A general phyletic dendro-gram, indicating these various anatid subgroups and illustrating representatives of each genus, is shown in Fig. 3. This diagram may be helpful for visualizing these broad intergroup relationships, approximately as they were initially illustrated by Johnsgard (1968b) and are still generally perceived.

What still remains as an obvious major taxonomic problem to be considered and discussed more fully is the sorting out of interspe-cific relationships within the genus *Oxyura,* for which the best cur-rently available evidence is provided by comparative behavior stud-ies. As such, this topic is best reserved for discussion in Chap. 4.

North American ruddy duck, male performing bubbling display to female.

C H A P T E R   T W O

Morphology, Anatomy, and Plumages

GENERAL MORPHOLOGY

To a degree greater than most of the waterfowl subgroups, the stiff-tails are rather readily characterized anatomically and morphologically. In the typical stifftails the tail is not only long, the feathers are numerous. There are usually 18 (sometimes 16) rectrices (tail feathers) in *Oxyura*; in *Biziura* there are 24, and in *Heteronetta* there are usually 14. By comparison, in the dabbling ducks there is a wide range, from 12 in the smaller species to as many as 20 in the larger ones (the freckled duck has 14). The individual rectrices are of graded lengths (Fig. 4B, C) and tend to be pointed (as descriptively indicated by the generic name *Oxyura*). The rectrices are also rather distinctly stiffened as a result of their enlarged shafts, which are slightly channeled below (as descriptively indicated by the older generic name *Erismatura*). These tail-feather adaptations in the typical stifftails certainly relate to the function of the tail in underwater maneuvering; a somewhat similar condition exists in the lengthened and stiffened rectrices of the torrent duck *(Merganetta)*. However, in this stream-dwelling species the feathers seem to be especially functional as a woodpeckerlike prop that is used when the birds are standing on slippery rocks; as in the stifftails, the rectrices of the *Merganetta* may secondarily serve as a male display structure in tail cocking. Except in the black-headed ducks, in which the tail coverts are nearly as long as the rectrices themselves, the tail coverts of stifftails are relatively short.

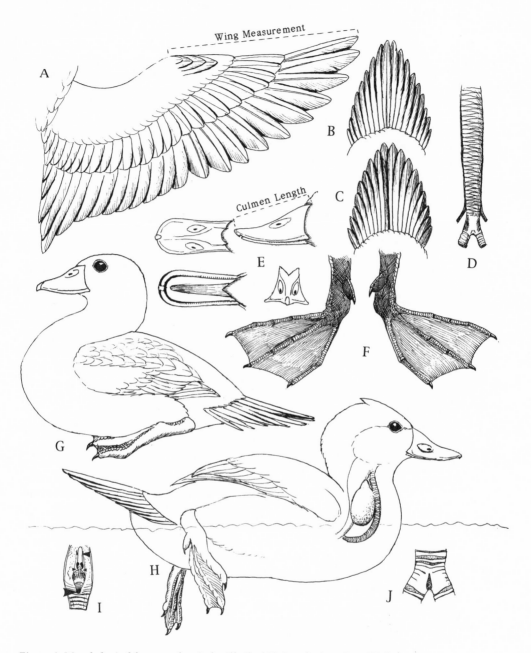

Figure 4. Morphological features of typical stifftails: (A) dorsal wing view, (B) dorsal tail view, (C) ventral tail view, (D) ventral view of male Australian blue-billed duck's trachea, (E) various bill profiles, and (F) foot characteristics. Profiles of (G) resting masked duck and (H) swimming ruddy duck, the latter showing an inflated tracheal air sac. (I) Ventral view of a male North American ruddy duck's larynx, showing the opening of the tracheal air sac (large pointer), the glottis (medium pointer), and the pulvini laryngis (small pointers). (After Wetmore 1918.) (J) Ventral view of a male musk duck's syrinx. (After Forbes 1882.)

The legs of stifftails are short, but their tarsi are covered by vertically aligned (scutellated) scales (Fig. 4F). The feet of typical *Oxyura* and *Biziura* species resemble those of many diving ducks in general shape, in that the hind toe is strongly lobed and the length of the other toe is nearly as great (ranging from 95 to 99 percent) as that of the middle toe (Cramp & Simmons 1977; Marchant & Higgins 1990). Relative lengths of the inner and hind toes are seemingly not very consistent with regard to diving abilities among ducks, but as with other diving ducks such as the pochards and sea ducks the hind toes of all stifftails other than *Heteronetta* are strongly lobed (Fig. 4F). The increased length of the outer toe probably helps increase the surface area of the foot and must thereby facilitate diving. By comparison, in *Stictonetta* the outer toe is about 93 percent as long as the middle one, and the hind toe is not strongly lobed. In typical *Anas* species the overall average is about 90 percent, but with considerable interspecies variability, and hind-toe lobing is slight.

The wings of the typical stifftails are relatively short and their bones weakly developed—in *Biziura* the wing-length to tail-length ratio is about 2.0:1, in *Oxyura* it is 2.4:1, and in *Heteronetta* it is 3.4:1. The wing-length to tarsus-length ratio in *Oxyura* and *Biziura* is likewise low, averaging about 4.3:1. By comparison, in *Heteronetta* and in *Stictonetta* it is about 5.0:1. As in nearly all other species of waterfowl there are 11 primaries, only 10 of which are functional, the outermost eleventh primary being rudimentary and minute. The wings are slightly rounded in shape when spread, with the ninth primary usually the longest, although the tenth is only slightly shorter (Fig. 4A). There are only 12 secondaries in typical *Oxyura* and *Biziura*, the fewest of any species of Anatidae. (The number in *Nomonyx* is not yet determined.) This low number of secondaries is related to the short wings of typical stifftails and may be compared with 17 secondaries in *Stictonetta* and 14 to 17 in most species of *Anas*, as well as in most pochards and sea ducks. A smaller number of secondaries may occur in a few semiflightless forms of *Anas* in which the wings have been greatly reduced (Marchant and Higgins 1990). Only one species of stifftail, the masked duck, has a distinctive color pattern, or speculum, on the secondaries. As in virtually all other waterfowl, several of the inner secondaries are somewhat elongated and pointed, forming distinctively shaped "tertials" that nearly, sometimes entirely, cover the tips of the primaries when the wing is folded.

The bill of typical *Oxyura* stifftails is distinctively shaped (Fig. 4E), being about as long at the head, and little higher than broad at the base, but greatly depressed and widened toward the tip, which is almost spatulate. The nail as viewed from in front or above is very narrow, but it is widened and slightly decurved, in a unique manner, when seen from below. In the masked duck the bill is not nearly so strongly depressed; the nail is as wide as in other ducks and is not recurved below. In several *Oxyura* species (especially the white-headed duck, less obviously the maccoa duck) the bill is greatly swollen at its base, especially in adults, which has been presumed (but not proven) to be correlated with relative brackish-water consumption and corresponding development of the salt gland for extrarenal disposal of excess salts in body fluids (Johnsgard 1965a). The entire world white-headed duck population, for example, now largely winters on a single lake in Turkey that is too saline to freeze during winter (Green and Anstey 1992). Perhaps the distinctive anterior-pointed and somewhat downward-oriented nostrils of *Oxyura* are also related to this same attribute, and perhaps this configuration is helpful in draining salty fluids from the bill. The nostrils of the musk duck are also somewhat forward-directed, and their nares are very large, but an early description of the skull anatomy (Pyecraft 1906) suggests that their "lacrymal glands" may be quite small and perhaps not well adapted for consumption of saltwater. However, the birds sometimes occur on brackish water during winter, and one of the present authors has noticed that an adult female removed from such a brackish location had water dripping occasionally out of her nostrils.

Interestingly, the bills of males of all *Oxyura* species uniquely turn a blue color prior to the breeding season; similarly, the tip of the black-headed duck's bill also becomes more bluish at that time. At least in the North American ruddy duck, and presumably in the other species as well, this sky-blue color is the result of structural changes that cause light-scattering rather than the result of a change in pigmentation (Hays and Habermann 1969). Males of the black-headed duck also have grayish-blue bills, except for a pink-to-reddish base. Unlike the situation in *Oxyura*, this bluish color does not vary seasonally in intensity, but it is probably structurally dependent, as preserved specimens lose their color and become a uniformly dull gray.

By comparison, the bill of *Biziura* is grayish-black, much heavier basally and not depressed at the tip, which produces a less concave culmen profile and a more massive overall appearance. The nail shape of *Biziura* is broad and dissimilar to that of *Oxyura*, being more like that found in typical ducks. In *Heteronetta* the bill is more *Anas* like overall, the nail being broad and not recurved and the tip being only moderately flattened. However, in *Stictonetta* the bill shape is surprisingly *Oxyura*-like, with a narrow and high base, a flattened tip, and a correspondingly concave culmen profile. Similarly, the nostrils open toward the front. However, the nail is relatively prominent and wide, neither distinctly narrowed above nor recurved below.

The necks of the stifftails are distinctively short (Fig. 4G, H), with 16 cervical vertebrae present in *Oxyura*, *Biziura*, and *Heteronetta*, as compared with 16 to 19 in the genus *Anas* (Woolfenden 1961). A stifftail's neck is also relatively thick-skinned, and males of several species have a well-developed layer of subdermal muscles that seem to be related to the inflatability of neck structures during sexual display (Wetmore 1917).

TRACHEAL AND SYRINGEAL ANATOMY

The tracheal and syringeal anatomy of male stifftails is of special interest. Functional variations relative to vocalizations will be considered in a later chapter; however, brief summaries are warranted here as a means of further defining the group anatomically. A general discussion of waterfowl tracheal anatomy and associated taxonomic variations has been published earlier (Johnsgard 1961b), and it will only be noted here that the stifftails differ from nearly all true anatine ducks in that males lack asymmetrically inflated tracheal enlargements (bullae). Instead many, if not all, of them have highly inflatable tracheal air sacs, throat pouches, or an inflatable esophagus.

A good illustration of the trachea of the male freckled duck does exist (Campbell 1899; redrawn by Johnsgard 1961b). On the basis of that drawing the freckled duck appears to have a syrinx that is very gooselike, the syrinx being both symmetrical and seemingly primitive in lacking obvious tympaniform membranes. The male does have two tracheal enlargements, located about a third and again about halfway down the length of the trachea, which are compressed laterally so that they are not conspicuous when the trachea is viewed from the front. No neck inflation occurs during male display in this species.

Phillips (1922–26) commented that the black-headed duck has the same full and loose neck skin as do ducks of the genus *Oxyura* and has a relatively simple trachea that lacks a bony bulla. The middle third of the trachea is soft, swollen, and about double the diameter of the other two thirds. The masked duck has some clear similarities to the stifftails in its tracheal and esophageal structures. The trachea of males is simple in structure and lacks a bulla. The middle of the tracheal tube is broad and thin-walled. On either side of the mouth there are slits forming the entrances to a thin-walled cheek pouch that extends back about 25 millimeters, just above the hyoidean muscles. This sac is evidently inflatable and responsible for the enlarged throat area of displaying males. The anterior end of the esophagus may also be inflatable (Wetmore 1926).

The male syrinx of *Biziura* was illustrated by Forbes (1882) and Beddard (1898) and has been redrawn here (Fig. 4J). Like that of *Oxyura* it is simple and symmetrical in structure; the tracheal tube is evidently not modified in any apparent manner but is slightly narrowed toward the junction of the bronchi. The syrinx consists of several (about five) fused tracheal and bronchial rings ossified to form a simple symmetrical structure very much like that of *Oxyura*. The esophagus of *Biziura* is evidently not inflatable, but there is an inflatable pouch below the tongue in males, which connects to the base of the mouth via a small circular aperture about the size of a pea. The pouch below it extends back to the base of the tongue and forward to the posterior end of the subgular wattle but does not extend into it (Forbes 1882). When filled with air, the laryngeal pouch produces

a substantial enlargement of the pharynx and may perhaps additionally influence vocal sound production.

The tracheal and esophageal situation in the male masked duck
is clearly much like that of *Oxyura*, based on a description by Wetmore (1918). He stated that the skin over the neck and upper breast
is loose, filled with fatty tissue, and "full as it is in the ruddy duck."
The trachea is simple, and the syrinx lacks an enlarged bulla. The tracheal tube is enlarged proximally, and on the ventral surface near the
upper end there is a small aperture that opens into a rounded sac
that has a diameter of about a centimeter when inflated. There is also
a somewhat larger extension that opens from a second aperture on
the dorsal surface of the trachea. Additionally, the esophagus can be
inflated with air, having then an elongated sac near its center that is
2.5 centimeters in diameter.

In the male North American ruddy duck the esophagus is not
noticeably modified for display, nor is the syrinx. The latter consists
of about five or six united tracheal and bronchial rings of which the
posterior two or three are split. The tracheal tube has a saclike extension on the dorsal surface near the anterior end (Fig. 4H), which
even among still partially downy birds can be inflated to a diameter
of about 20 millimeters (Roberts 1932). Among adult males it is capable of much greater expansion and typically assumes a pearlike
shape, being about 50 to 65 millimeters long and 32 to 43 millimeters
deep and having a narrow anterior extension about 8 to 10 millimeters long. The mouth of the air sac lies on the dorsal surface of the trachea, immediately behind the larynx. The opening (Fig. 4I, large
dagger) has abrupt sides and is slightly longer than it is wide, with
a roughly triangular shape (Wetmore 1917). Inflation of the tracheal
air sac is very gradual; once filled, it is very resistant to deflation from
external pressure (Wetmore 1918).

The paired muscles (the sternotrachealis according to Wetmore,
more probably the ypsilotrachealis) that arise from the thyroid carti-
lages pass below the anterior end of the air sac where each of the two
muscles there expands to about 10 millimeters in width, but again
gradually narrows to about 2 millimeters wide at the posterior margin of the air sac. Contractions of these muscles probably presses the
air sac against the esophagus and closes the air-sac aperture, keeping
it inflated (Wetmore 1918). Initial inflation of the sac may be achieved
by passing air forward from the true pulmonary air sacs of the lungs.
The anterior glottal opening of the trachea is held shut by special pads
(pulvini laryngis, small paired daggers in Fig. 4I) and a median flap
(lingula laryngis, medium-sized dagger in Fig. 4I) that is functionally comparable to the mammalian epiglottis in closing the opening
of the trachea. All of these special features are essentially lacking in
female ruddies (Wetmore 1917).

In addition to these tracheal and laryngeal specializations of male
ruddy ducks, there are also paired dermal muscles that arise on each

side of the head behind the eyes and pass backward, converging below the larynx until they nearly meet. There they expand to cover all of the loose skin on the ventral sides of the neck and finally insert on the furculum (wishbone) below the neck (Wetmore 1917). These muscles probably also participate in neck inflation and contraction; perhaps they are significant with regard to the drumming action of the bill on the neck when the male is actively performing his "bubble" display.

The male tracheal anatomies of nearly all of the other species of *Oxyura* remain to be described, although nearly all of them are evidently capable of neck inflation by some means. This inflation (and associated neck-feather erection) is especially apparent in the maccoa, the Argentina blue-billed duck, and the Australian blue-billed duck. In the male Argentine blue-billed duck there is a long, narrow tracheal air sac measuring about 10 by 65 millimeters, the lingula laryngis is reduced, and there is no pulvini laryngis However, the esophagus is enlarged and inflatable (Wetmore 1926). In the Australian blue-billed duck a tracheal air sac is seemingly altogether lacking, but the esophagus can been inflated (Johnsgard, pers. obs.). Neck inflation is scarcely evident if present at all in the white-headed duck. It is also much less apparent in the Peruvian race of the ruddy duck than in the North American race; the Andean race would seem to be somewhat intermediate.

STERNUM AND APPENDAGE ANATOMY

In *Heteronetta* both the sternum and the closely associated coracoid bone have attributes approaching the condition found in *Anas* (Woolfenden 1961). In this genus the sternum is the narrowest of any species of the tribe, being only 43.9 percent of the sternum's basin length and 39 percent of the total sternum length (Woolfenden 1961; Raikow, 1971).

Whereas in nearly all other waterfowl a fairly large and oval or round pneumatic foramen is present on the dorsal side of the sternum near its anterior end, in all the stifftails the pneumatic foramen is either minute or entirely absent (Woolfenden 1961). Perhaps the absence of this feature relates to the sternum's poorly developed keel among stifftails. The sternum of the typical stiff-tailed ducks is relatively shallow—and very wide relative to its length. The medial width of the sternum relative to its basin length ranges from 59.5 to 68.4 percent in typical *Oxyura*; in the masked duck, it is 48 to 54.1 percent. By comparison, in *Anas* the width of the sternum varies from 37.5 to 48.9 percent of its total length (Woolfenden 1961).

As just noted, the sternum of typical stifftails has an extremely low keel that reflects the rather weak flying ability of these birds. In *Biziura* the keel is even more poorly developed than it is in the poorly flying to virtually flightless steamer ducks, *Tachyeres*, and, like the condition in the steamer ducks, the wing bones of the musk duck are relatively slender and the phalanges are reduced (Pyecraft 1906). The sternum of the musk duck is very wide, its width being 58 percent of the total sternum length (Raikow 1971). However, because the keel in *Biziura* is so greatly reduced—and because of the reduction in area posterior to the costal margin associated with the reduced keel—the sternum's width is only 41 percent of its basin length (Woolfenden 1961). Interestingly, the musk duck has an extremely large number of ribs (ten thoracic ribs, of which eight reach the sternum), more than in any other duck and equal in number to a few swans. In *Oxyura* there are nine thoracic ribs, of which eight are attached. This unusually large number of ribs in stifftails was considered a primitive trait by Pyecraft (1906), but perhaps it also relates to diving adaptations associated with general strengthening of the rib cage.

The pelvic girdle consists of the largely fused ilium, ischium, and pubis bones, which collectively with associated vertebral elements form a common pelvis. In the stifftails the pelvis is particularly narrow, which maximizes body streamlining but also makes efficient land locomotion more difficult since the legs are placed so close together and are situated far behind the bird's center of gravity. The interacetabular pelvic widths, measured as the distance between the dorsal edges of the two acetabula, exhibit a progressive narrowing in the sequence *Anas, Heteronetta, Oxyura,* and *Biziura* (Fig. 5A). The narrowest pelvis of any anatid species is found in *Biziura,* where the interacetabular width is 14.7 percent of total length. That of the masked duck averages only slightly wider (19.91 percent) than the mean pelvis width of the other *Oxyura* species (18.96 percent) (Woolfenden 1961). In *Heteronetta* the corresponding ratio is 23 percent, and in a representative species of *Anas* (the mallard) it exceeds 30 percent (Raikow 1971). Such pelvic narrowing permits the heads of the femurs to rise more medially from the pelvis, reducing the cross-sectional area of the bird and thus minimizing water resistance (Raikow 1970). Additionally, the postacetabular extension of the pelvis relative to overall pelvic length is remarkable among stifftails, with its relative length increasing in the sequence *Anas, Heteronetta, Oxyura,* and *Biziura* (Fig. 5B). The postacetabular elongation in *Biziura* (68 percent) is not only the greatest of all the Anatidae, it is also greater than that occurring in such diving specialists as loons and grebes; as noted below, the resulting proportions strongly influence the effective angle of action of associated femoral muscles (Raikow 1970).

As noted earlier, the feet of stifftails are relatively large, and their femurs and tarsi are relatively short. Together, these traits allow for a powerful foot stroke and a maximum amount of water interception during foot paddling. The ratio of the middle-toe length to the length of the femur is about 0.9:1 for *Anas,* about 1.15:1 for *Heteronetta,* and about 1.25 to 1.3:1 for *Oxyura* and *Biziura* (Raikow 1970). Additionally, the middle-toe length to tarsus length ratio in *Oxyura* and *Biziura* averages about 1.8:1. This ratio is about 1.7:1 for *Heteronetta,* for

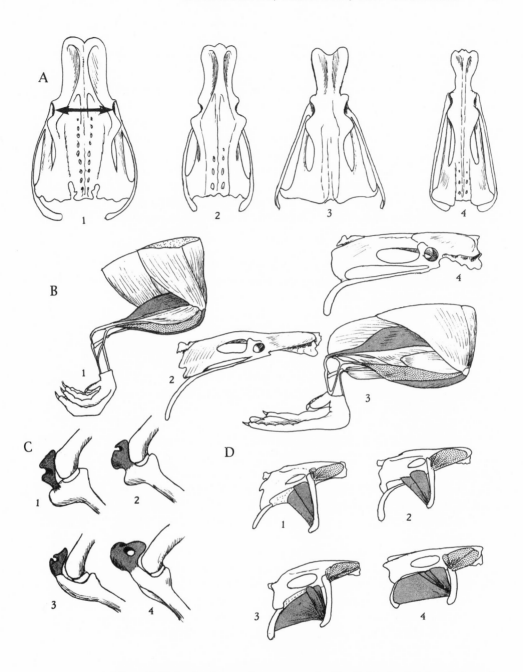

Figure 5. Skeletal and muscular anatomy of stiff-tailed ducks: (A) dorsal views of pelvic bones of (1) Anas, (2) Heteronetta, (3) Oxyura and (4) Biziura. Arrows indicate interacetabular distances. (B) Lateral views of pelvic area and associated musculature of (1, 2) Anas and (3, 4) Biziura. The gastrocnemius is shown by fine stippling, the peroneus longus by coarse stippling. (C) Knee area (patella stippled) and (D) deep thigh muscles (extensors shown in fine stippling; flexors in coarse stippling) of (1) Anas, (2) Heteronetta, (3) Oxyura, and (4) Biziura. (After drawings in Raikow 1970.)

Stictonetta it averages about 1.3:1, and among typical *Anas* species it ranges from about 1.25 to 1.4:1 (Cramp and Simmons 1977; Marchant and Higgins, 1990).

The femur of stifftails, like that of other diving ducks, tends to be quite robust and short; it is dorsoventrally curved, with well-developed scars for extensive muscle attachments. In *Heteronetta*, of all the stifftails, it is least modified for diving, and its shaft is the thinnest (6.9 percent) of all three stifftail genera. In the masked duck it is also little specialized, but in the typical *Oxyura* its shaft thickness ranges from 8.1 to 8.5 percent of its length. In *Biziura* it is the most robust, with a shaft thickness of 10.1 percent of its length (Woolfenden 1961). The combination of a relatively short femur in the stifftails, in addition to a much longer postacetabular component of the pelvis, results in a considerable shift in the functional working angles of muscles (especially the adductor longus et brevis) that produce extension (rearward movement) of the femur, which in turn results in a much more directly horizontal line of power action (Fig. 5D). The trend reaches its extreme in *Oxyura* and *Biziura*, in which genera this muscle is more highly modified than is true of any other muscle of the hind limb. Extension (backward movement) of the femur is certainly of considerable importance for land walking. It may also be important during aquatic locomotion by counteracting the tendency for the femur to become flexed when the tarsus is extended during the power stroke (Raikow 1970).

The middle leg element, the tibiotarsus, is also variably robust relative to the diving abilities of each species. In *Heteronetta* the shaft thickness is 4.1 percent of its length; in the *Oxyura* species it is 5.0 to 5.4 percent; and in *Biziura* it is 6.4 percent (Woolfenden 1961). The functional length of the tibiotarsus is increased by a fusion of the inner cnemial crest of the tibotarsus and the patella (Fig. 5C). In *Anas* and *Heteronetta* the inner cnemial crest of the tibiotarsus is hatchet-shaped and relatively short, but in *Oxyura* and *Biziura* it is shallow, elongated, and fused with the patella. In *Anas* the patella is mostly ligamentous, but in the stifftails it is highly ossified. In *Biziura* the patella is so large that the groove for the ambiens tendon has become entirely enclosed to form a foramen, in a manner unique to anatids but similar to the condition found in cormorants, which are also diving specialists. In *Biziura* the total elongation of the tibiotarsus anterior to its articular surface by these structures is 43.73 percent, as compared with 30.95 percent in *Oxyura* and 23.5 percent in *Heteronetta*. These proportional trends reflect the relative specializations in these genera for underwater locomotion (Raikow 1970).

The tarsus (tarsometatarsus) of stifftails is also relatively wide and robust, its shaft width ranging from 10.8 to 16.3 percent of its length. In correlation with relative diving abilities, *Heteronetta* has the proportionally narrowest shaft (10.8 percent) and *Biziura* the stockiest (16.3 percent), with *Oxyura* (including *Nomonyx*) intermediate

(11.6 to 12.8 percent). Among dabbling ducks this ratio varies from 8.3 to 11 percent (Woolfenden 1961).

During resting and leisure swimming, the legs of *Oxyura* species are oriented generally downwards (Fig. 4H). With stronger swimming, and especially during underwater locomotion, the axis of the tarsus is raised to about the level of the horizontal, or perhaps even beyond the horizontal while the birds are progressing underwater. (See Fig 7.) Swimming speed and power are generated by an extension of the tarsus through the contraction of two large muscles, the gastrocnemius and the peroneus longus. In *Oxyura* and *Biziura* the belly portions of both these two muscles are relatively long, but the pars interna of the gastrocnemius is especially increased in mass relative to *Anas* species (Fig. 5C). Raikow (1970) estimated that during contraction this muscle of *Biziura* produces 71 percent more force per unit of contraction than occurs in mallards, as compared with 37 percent in *Oxyura* and 15 percent in *Heteronetta*. The power stroke is followed by and alternated with a recovery stroke, or flexion, which is produced by a contraction of the tibialis anterior muscle. The strength of the recovery stroke of the tarsus is roughly indicated by the relative size of this muscle; this has been calculated as providing about 13 percent greater force of contraction in *Oxyura* than occurs in *Anas* (Raikow 1970).

Comparative data on foot surface area relative to body mass in the stifftails are not yet available, but such data might prove instructive as an estimate of surface area available for propelling the body through the water. As a minimum comparison, surface-area estimates for two mounted male ruddy duck feet (excluding their hind toes) averaged 1866 square millimeters (1850 and 1882 square millimeters). This estimate represents an average of about 3.4 square millimeters of flat surface available from each individual foot for propelling each gram of body mass, assuming an average adult male weight of 550 grams. Similar surface-area estimates for two male mallard feet averaged 1985 square millimeters, representing 1.6 square millimeters of foot area available for propelling each gram of body mass, assuming an average adult weight of 1200 grams. This rough index would suggest that a ruddy duck might have about twice as much foot propellant area available per unit of body mass than has a mallard; the former might thus be a correspondingly more efficient swimmer and diver. As to comparisons with a few other more typical diving ducks, for a male common merganser *(Merqus merganser)* the estimated ratio was 2.2 square millimeter per gram, for a male oldsquaw *(Clangula hyemalis)* it was 2.3 square millimeters per gram, and for a female bufflehead *(Bucephala albeola)* it was 4.6 square millimeter per gram. These few numbers, which should be substantiated using live birds, would suggest that a female bufflehead might a substantially better diver than the common merganser, and perhaps even better than a ruddy duck. However, they might also indicate that smaller species

TABLE 2
Terminology for Molts and Plumages

Usage in Humphrey and Clark (1964)		General Usage in World Literature and in Present Text	
Plumage	Molt	Plumage	Molt
	—	Natal	
Juvenal			Postnatal
	Prebasic	Juvenal*	
Basic I			Postjuvenal
	Prealternate I	First nonbreeding†	
Alternate I			First prebreeding
	Prebasic II	First breeding‡	
Basic II			First postbreeding
	Prealternate II	Second nonbreeding†	
Alternate II			Second prebreeding‡
	Prebasic II	Second breeding	
			Second postbreeding

*Juvenal here refers to the plumage, juvenile to the age class. Some authorities use juvenile in referring to molts as well.

†Nonbreeding duck plumages are also often called nonnuptial or, less often, winter plumages, the latter especially if they are carried through the entire winter season. In many duck species the male nonbreeding plumages are femalelike and are typically acquired in late summer but lost by early winter. These nonbreeding plumages are often called eclipse plumages, rather than winter plumages.

‡When a plumage no longer changes in pattern or color with increasing age it is often called an adult or definitive plumage. In stifftails the first definitive breeding plumage may develop during the first or second year after hatching, although sexual maturity may not always correspond with the initial assumption of this plumage. Sex-related differences may also exist in acquisition of definitive plumages and in attaining sexual maturity.

of diving ducks in general are likely to have more favorable area-to-mass ratios than do larger ones.

MOLTS AND PLUMAGES

One of the basic features used by Delacour and Mayr (1945) in distinguishing the true ducks (Anatinae) from the geese, swans, and whistling ducks was that in the former subfamily there are typically two body molts, and thus two recognizable plumages, per year. In nearly all species of both subfamilies a single, nearly simultaneous, molt of flight feathers (primaries and secondaries) is considered typical; this molt occurs shortly after breeding and produces a flightless period of variable length. However, the tail feathers (rectrices) may be molted once or twice per year in ducks in conjunction with the body molt, the feathers usually being dropped and replaced in graduated and often irregular sequence.

Until recently, very little was known of the molts and plumages of the stifftails. However, as was summarized by Humphrey and

Clark (1964), at least the North American ruddy duck was known to retain its juvenal plumage until midwinter. At that time the young birds typically molt most or all of their feathers and replace them with a second immature plumage ("basic I" plumage of Humphrey and Clark) that is carried until spring. [Note that Hughes (1990) found that this postjuvenal molt did not occur among 12 captive birds that he studied at the Wildfowl Trust.] This postjuvenal molt may be unique in the Anatidae in that it is apparently complete, involving not only the body feathers, but also the flight feathers (remiges) and tail feathers (rectrices). This plumage in turn is lost by a molt that brings the yearling bird into a plumage resembling the adult breeding plumage ("alternate plumage" of Humphrey and Clark). Thereafter, ruddy ducks have an annual molt cycle consisting of complete postnuptial ("prebasic") molts in the fall and complete prenuptial ("prealternate") molts in spring.

For convenient reference, the terminology of molts and plumages used by Humphrey and Clark (1964), as compared with that generally used in world literature (e.g., Cramp and Simmons 1977; Marchant and Higgins 1990)—as well as in this book—may be summarized as shown in Table 2.

Since the general summary by Humphrey and Clark, little additional detailed information has accrued on comparative molts and plumages in the stifftails. However, a fair amount of information has accumulated on individual species. Such details of molts as well as plumage descriptions can be found in the individual species accounts.

North American ruddy duck, female with brood of four ducklings.

General Behavior and Ecology

POSTURAL AND LOCOMOTORY BEHAVIORS

Most egocentric (self-directed) activities of stifftails, as well as all of their social interactions, are much more likely to occur on water than on land, since most stifftail species spend very little time on land. When standing or walking, typical stifftails assume a much more erect position than do more terrestrially adapted ducks (Figs. 6A, B; 7F); the stifftails' characteristic posture is required to bring their center of gravity back to the axis of the legs and feet. However, when a stifftail is resting or sleeping on land the breast is brought down into contact with the substrate, and stifftail ducklings usually, if not always, rest on their bellies while on land (Fig. 6C). However, in contrast to assertions by Kortright (1943), this is not the normal terrestrial posture of adults, nor are North American ruddy ducks forced to progress on land by laboriously pushing themselves forward by using both feet simultaneously, in the manner typical of loons and grebes.

The black-headed duck maintains a distinctly horizontal body posture when standing on land, a posture associated with more forwardly placed and more widely spaced legs (Fig. 7D). This terrestrial posture is not very different from that characteristic of freckled ducks on land (Fig. 7A), nor indeed of typical dabbling ducks *(Anas)* and other nondiving ducks generally. Compared with stifftails, these groups have relatively longer legs, allowing for more rapid walking. Their legs are also more widely spaced because of their wider pelvic

Figure 6. Characteristic terrestrial and aquatic postures of typical stifftails: (A, B) adults standing on land, (C) duckling resting on land, and (D–G) adults resting or sleeping on water. (After photos by Johnsgard.)

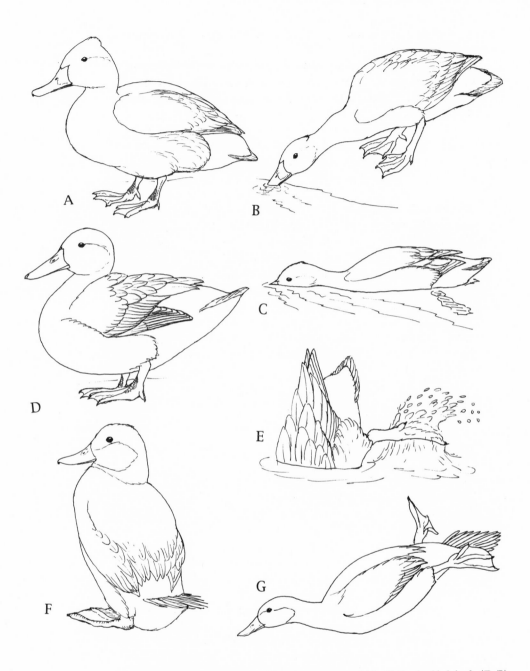

Figure 7. Typical standing and foraging postures of (A–C) the freckled duck, (D, E) the black-headed duck, and (F, G) the North American ruddy duck. [After photos by Johnsgard, (A–F) and in Todd 1979 (G).]

girdles, which probably produces a stable body support and results in a distinctly waddling gait.

Accompanying different leg placements are marked differences in diving abilities and foraging tendencies among these representative body types. Freckled ducks normally feed at the water surface, most often by bottom-filtering with the tip of the bill while standing on land at the water's edge or in very shallow water (Fig. 7B). Freckled ducks sometimes also forage by surface-filtering while swimming in deeper water, with the bill fully submerged and the eyes held close to the water (Fig. 7C). Foraging dives have not been reported in freckled ducks (Frith 1965).

Black-headed ducks very often forage by surface-filtering in a fashion similar to that typical of freckled ducks (see Fig. 14) and occasionally used by typical stifftails. However, like freckled ducks and typical dabbling ducks, but unlike typical stifftails, black-headed ducks sometimes "tip up" ("upend" in British terminology) to reach bottom or near-bottom foods in shallow waters (Fig. 7E). Diving may also be done by black-headed ducks in waters slightly deeper than can be reached by tipping up. However, as with typical dabbling ducks, diving by black-headed ducks is performed with a distinct upward jumping motion and with considerable splashing. Diving is thus done not nearly so gracefully and easily as it is by typical stifftail species of *Oxyura* (Fig. 7G) or by the musk duck, both of which can smoothly slip below water quietly and without splashing, in a manner similar to that of grebes.

MAINTENANCE, COMFORT, AND QUASI-SOCIAL BEHAVIORS

Maintenance behaviors include all those fundamental behavioral activities critical to maintaining an individual's life processes, such as ingestion of food, egestion or elimination of solid or liquid wastes, drinking, maintenance of body temperature and internal homeostasis, resting, and sleeping. Resting, either on land or in the water, is done with the head turned back and the bill tucked into the scapulars (Fig. 6D–G). Such resting behavior is often not distinguishable from sleeping and may be interspersed occasionally with other low-level activities. This posture, called "pseudosleeping," is common in all stifftails, but there is no good evidence that it serves as an attack inhibitor, as has been alleged (Palmer 1976). Drinking by stifftails, as well as by all other waterfowl, is performed by a bill-dipping movement, followed by a rapid return of the bill to a level somewhat above the horizontal.

Beyond these basic life-sustaining activities, there are additional "comfort movements" (McKinney 1965) that may not be directly required for maintaining life, but which perhaps help sustain its quality. Many of these self-directed activities involve care of the integument and the associated feather coat, such as scratching various areas of the head region with the middle toenail while swimming or standing on one leg. Feather preening (Fig. 8B, C; also 9C, E) and the associated removal of loose feathers with the bill ("nibbling-preening") are frequent activities during molting periods. Tail-fanning and wing-shuffling movements are common during preening sessions. Wing flapping (Fig. 8G) is common throughout the year, as is a general overall plumage shaking, which may be performed either on land ("body shake," Fig. 8H) or in the water ("swimming shake," Figs. 8J and 9F). Both of these activities may serve to rearrange feathers or to help remove water from them. More restricted shaking movements, such as wing-shaking, tail-shaking ("tail wagging"), and foot-shaking movements, also often occur; water or mud is usually removed from the bill or feet with quick lateral head shaking or by more vertically oriented head flicking (Fig. 8K).

All waterfowl obtain their uropygial-gland secretions ("oiling-preening") by nibbling at the duct (Fig. 8A) or by rubbing the head over the duct and then spreading the oil over the plumage with the bill and head (Fig. 9D). Muscle-stretching movements include a yawning-like "bill stretching" and two methods for stretching the wings: namely, stretching each wing individually and horizontally backwards while the corresponding leg is simultaneously stretched backwards in a "wing-and-leg stretch" (Figs. 8E, 9H), or stretching both wings simultaneously and vertically in a "two-wing stretch" (Fig. 8F). Finally, aquatic bathing is common and includes repeated head dipping, splashing water over the back (Fig. 8M), energetic wing thrashing, or a more acrobatic "somersaulting" (Fig. 8I). Bathing is usually followed by a bout of prolonged preening, and both activities occur during apparently leisure-time periods. A few other miscellaneous comfort movements, such as foot or leg nibbling and bill dipping (Fig. 8L), also occur among stifftails and probably all other waterfowl species.

McKinney (1965) reported observing typical stifftails (specifically, North American ruddy ducks) performing the general body shake (on land and in shallow water), the swimming shake (in deeper water), head shaking, head flicking, tail wagging (here called tail shaking), wing flapping, foot shaking, and wing shuffling plus tail fanning, wing-and-leg stretching, two-wing stretching, jaw stretching, scratching, and bill dipping. He also reported seeing typical oiling-preening behavior (including the head-rolling and cheek-rubbing components of oil spreading), nibbling-preening behavior, and several bathing components (head dipping, wing thrashing, and somersaulting). During his studies of the black-headed duck, Weller (1968a) observed all the comfort movements described by McKinney for anatids in general.

Social mutual preening (allopreening) has not yet been reported for any true stifftails, but it has been observed in the black-headed duck (Rees and Hillgarth, 1984). Though McKinney did not personally

Figure 8. Comfort movements of various Anatinae: (A) oiling-preening, Baikal teal; (B, C) nibbling-preening, ringed teal; (D) jaw stretching, surf scoter; (E) wing-and-leg stretch, Laysan teal; (F) two-wing stretch, bufflehead; (G) wing flapping, bufflehead; and (H) general (standing) body shake, common merganser. Also shown are (I) somersaulting-bathing sequence, (J) swimming shake, (K) head flicking, (L) bill dipping, and (M) head dipping, all in Anas. [After photos by Johnsgard (A–H) and sketches by McKinney 1965 (I–M).]

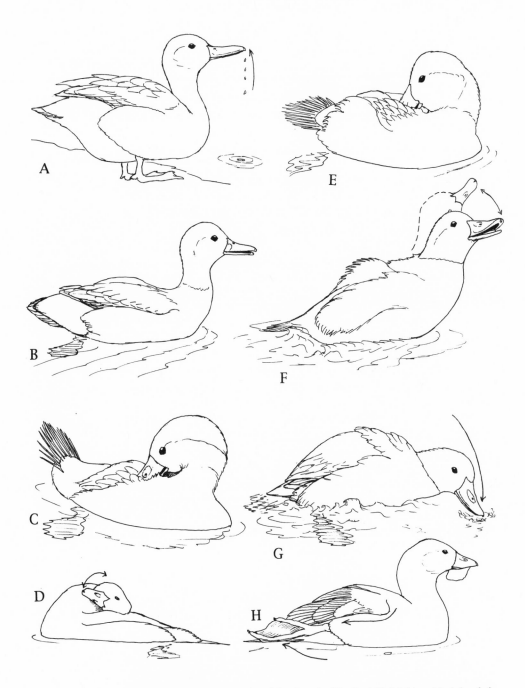

Figure 9. General and comfort behavior of stifftails: (A) drinking, black-headed duck; (B) preflight posture, black-headed duck; (C) nibbling-preening, musk duck; (D) cheek rubbing, Australian blue-billed duck; (E) nibbling-preening, Argentine blue-billed duck; (F) swimming shake, Argentine blue-billed duck; (G) bathing (head-dipping phase), maccoa duck; and (H) wing-and-leg stretch, musk duck. (After photos by Johnsgard.)

record any preflight movements for ruddy ducks, he noted that head shaking and possibly shoulder rubbing (rolling the cheeks on the scapulars) have been reported and may occur in this situation. Likewise, no regular preflight movements were observed by Johnsgard (1965a) or Carbonell (1983) in any of the typical stifftails. However, Hughes (pers. com.) has seen the birds sometimes perform head shaking while in an alert posture prior to taking flight. The similar alert, preflight posture of the black-headed duck is one in which the neck is vertically stretched and the bill is held horizontal (Fig. 9B). During a 13-hour observation period McKinney observed four cases of oiling-preening and a mean of 5.2 bathing episodes per individual in two North American ruddy ducks. At least at the Wildfowl Trust, a surprising amount of time is spent simply resting during daylight hours by most species in the summer months, with foraging and comfort behavior generally occupying progressively shorter periods of time. When preening their long tail feathers, musk ducks and other typical stifftails sometimes pivot or spin rapidly about in the water. However, this curious behavior is probably simply a reflection of the difficulties the birds have in reaching their long tails and is of no special social significance.

In common with many other duck groups, many waterfowl comfort movements have clearly been ritualized over time, coming to serve as important social display functions. Bill-dipping, head-dipping, wing-shuffling or wing-shaking movements, display preening, head shaking, aquatic body shaking (swimming shake), tail shaking, and similar actions all frequently occur during stifftail display. In some stifftail species these postures or actions may occur fairly regularly in a display context, but the associated movements are not always clearly distinct in form from those occurring in nonsocial "comfort-behavior" situations. A ritualized version of the swimming shake is shown for *Oxyura* in Fig. 9F, and the normal preening postures of musk duck and maccoa duck shown in Fig. 9C and E may likewise be compared with ritualized "dab-preening," as illustrated in the account of the Australian blue-billed duck.

The searching component of feeding, or foraging, is obviously related to individual maintenance behavior. However, it is better regarded as a separate behavioral category, inasmuch as different foraging adaptations in various species and different foraging depths achieved result in great differences in amounts of time actually spent in searching for and obtaining food and in associated energy expenditures. Stifftails do not normally forage socially, but individual foraging activities in favored sites may bring birds into proximity and thus facilitate subsequent social interactions between them. Foraging black-headed ducks at the Wildfowl Trust often swim side by side, and occasionally they follow one another about while surface-foraging. Perhaps they thereby benefit somewhat by obtaining foods that may be stirred up and brought to the surface by the birds swimming

TABLE 3
Breeding-Season Diurnal Activity Budgets of Adult Male Stifftails

	N*	Resting,† %	Feeding,† %	Comfort Behavior,† %	Swimming,† %	Sexual Display,† %
Black-headed duck	4189	42.4	19.9	15.2	14.5	6.3
North American ruddy duck	4083	46.2	18.3	12.2	15.8	7.1
Maccoa duck	7624	42.1	13.7	14.9	12.6	13.5
Argentine blue-billed duck	8643	34.4	27.0	15.8	14.9	5.5
White-headed duck	8563	40.2	13.6	19.2	12.4	14.5
Musk duck	2002	25.0	25.4	17.6	22.3	9.7

*N equals number of 30-second observation intervals in sample.
†Figures represent percentage of total activities tallied. Some minor activities were not tabulated here, so percentages may not total 100.
Source: Data obtained between March and August at the Wildfowl Trust (Carbonell 1983).

beside or directly ahead of them. Additionally, ruddy ducklings have been seen feeding on matter brought to the surface by the foraging actions of the brood female (Hughes, pers. com.).

Virtually all stifftail species dive not only to obtain food but also to escape from danger, although the black-headed duck is more prone to swim to the middle of a pond and maintain a vigilant watch from there. If further disturbed, this species is more likely to fly than dive (Weller 1968a), which is in strong contrast to the escape-diving characteristic of typical stifftails.

Tables 3, 4, and 5 summarize information on general activity categories for several species of stifftails as observed by Carbonell during her studies at the Wildfowl Trust. Somewhat more detailed breakdowns (by time of day) of these same data, as well as some related observations, are also available in unpublished form in Carbonell's (1983) original summary. These tables indicate significant sexual differences in mean daily activity patterns of adults during the breeding season, and even greater changes in temporal female activities at various stages of the breeding season. Generally there are few interspecific differences among the females at corresponding stages of breeding, and indeed rather surprising similarities among them. Duckling behaviors undergo considerable proportional changes in activity patterns during their first month of life, but feeding and resting typically occupy large parts of their daily activities.

Rather little is known of nocturnal behavior activities of stifftails. However, at least during the breeding season wild musk ducks can often be heard displaying during all hours, including those of total darkness (Johnsgard pers. obs.). Carbonell also obtained some nighttime observations of this species. She once determined that display

TABLE 4

Breeding-Season Diurnal Activity Budgets of Adult Female Stifftails

	N*	Resting,† %	Feeding,† %	Comfort Behavior,† %	Swimming,† %	In Nest† %
Prebreeding Period						
Black-headed duck	4189	43.0	26.4	16.1	12.9	0.0
North American ruddy duck	1588	46.5	28.9	11.3	10.1	0.0
Argentine blue-billed duck	3692	45.1	33.9	12.3	7.4	1.0
White-headed duck	2772	43.1	28.6	13.8	14.5	0.0
Prelaying Period						
North American ruddy duck	1386	29.7	47.5	11.5	10.1	1.0
Maccoa duck	861	20.4	38.8	13.4	17.9	4.4
Argentine blue-billed duck	1847	14.6	49.5	15.8	18.4	0.4
White-headed duck	2638	27.6	57.2	8.2	6.4	0.6
Egg-Laying Period						
North American ruddy duck	884	30.2	27.1	9.3	14.7	18.7
Maccoa duck	622	17.5	56.3	2.4	5.8	17.7
Argentine blue-billed duck	1818	16.5	35.1	7.2	6.3	34.3
White-headed duck	263	5.4	37.6	3.9	4.34	8.7
Incubation Period						
North American ruddy duck	11,028	—	4.2	1.6	0.8	92.4
Maccoa duck	12,369	—	5.8	2.1	1.1	91.0
Argentine blue-billed duck	13,572	—	4.3	1.5	0.8	93.3
White-headed duck	14,205	—	5.5	1.1	0.1	92.7
Leading Young, 1st Week						
North American ruddy duck	1870	15.8	11.4	11.9	35.3	22.2
Maccoa duck	4859	20.7	17.5	12.9	40.1	8.1
White-headed duck	3330	14.7	22.7	8.1	16.2	36.3
Leading Young, 3rd Week						
North American ruddy duck	515	33.8	11.2	27.4	27.6	0.0
Maccoa duck	1340	38.0	17.4	20.8	23.2	0.0
White-headed duck	1230	11.2	33.7	17.0	11.2	0.0

* N equals number of 30-second observation intervals in sample.
†Figures are shown as percentages of all activities tallied.
Source: Data obtained between March and September at the Wildfowl Trust (Carbonell 1983).

calls of one male began at 8:45 P.M. and continued until 9:30 P.M. Periodic calling occurred about 3:00 A.M. began again a half hour later, and continued until at least 4:00 A.M.

During one full night of observing black-headed ducks, Carbonell noted that a high level of activity (mostly displaying and feeding) occurred until about 8:30 P.M. Thereafter activities declined and most birds were asleep by midnight. After the moon began to rise at about 1:30 A.M., the birds began to feed quite actively.

During one night's observations of North American ruddy ducks, Carbonell found that the birds displayed until about 10:00

TABLE 5
Diurnal Activity Budgets of Young Stifftail Ducklings

	Age of Ducklings, Weeks*			
	1	2	3	3+
North American ruddy duck				
Resting	19.6%	46.4%	45.5%	—
Feeding	44.6%	41.1%	39.1%	—
Comfort behavior	9.0%	7.6%	12.6%	—
Swimming	16.2%	4.8%	2.8%	—
In nest	9.3%	—	—	—
Sample size†	8989	5823	2499	—
Maccoa duck				
Resting	30.6%	22.7%	46.4%	53.9%
Feeding	37.0%	54.1%	36.7%	31.5%
Comfort behavior	5.5%	7.0%	7.3%	7.1%
Swimming	18.0%	16.2%	9.6%	7.4%
In nest	8.8%	—	—	—
Sample size†	13,499	3847	1658	2688
White-headed duck				
Resting	21.0%	38.1%	38.0%	44.8%
Feeding	22.7%	50.1%	53.4%	42.9%
Comfort behavior	9.5%	7.0%	6.4%	9.5%
Swimming	16.2%	4.8%	2.2% -	2.8%
In nest	35.3%	—	—	—
Sample size†	10,501	3515	2716	4799

*Figures shown as percentages of total activities tallied.
†"Sample size" defined as in Tables 3 and 4.
Source: Data obtained at the Wildfowl Trust by Carbonell (1983).

P.M., when courting stopped and feeding began. At least one male and female were still actively preening and swimming at 10:15 P.M. The birds were also seen diving and preening at 3:00 A.M. In a night's observations of the Argentine blue-billed duck, Carbonell noted that social activities (mostly displays) continued until about 9:15 P.M., when feeding began. Feeding and displays continued until about 2:30 A.M. . Resting then occurred until about 3:30 A.M., when feeding began again. Feeding continued until at least 4:00 A.M., when observations ceased. Few nighttime observations of the white-headed duck were made, but these birds also were seen displaying until well after 9:30 P.M.

In general, Carbonell's observations indicated that most foraging by male stifftails occurred in the afternoon hours and that during the breeding season all females spent significantly more time foraging than did males, these activities being spread out over most of the day. All stifftail species tended to display more after 4:00 P.M.— except for the maccoa duck, which mostly displayed during early morning hours. A rather surprising amount of foraging and social display occurred during hours of darkness in all species closely observed by Carbonell. This nocturnal pattern of foraging by ruddies was later observed in wild ruddy ducks wintering on the Atlantic coast by Bergan (1986), who suggested that nocturnal activity patterns of a common midge species could account for such feeding patterns. That is, midge larvae move out of the deeper layers of the silty substrate and into the water column and thus become more available to diving ducks. Hughes (1992) found a similar, indeed even more pronounced, nocturnal foraging pattern among feral ruddy ducks in England and suggested the same cause. He was not certain whether this activity pattern represented a preferred foraging strategy or an obligatory adaptation. For varied ecological reasons, nocturnal feeding by waterfowl may indeed be fairly common (McNeil, Drapeau, and Goss-Custard 1992).

COMPARATIVE FORAGING BEHAVIORS AND FOODS

Of all the waterfowl groups, the stifftails are among the most isolated in their foraging behavioral ecologies. By virtue of their remarkable diving abilities and foraging adaptations, they have opportunities for foraging on underwater foods that are found on the silty bottoms of marshes and ponds and are relatively unavailable to other groups of waterfowl (Tome and Wrubleski 1988). They often locate their foods by tactile rather than visual clues—using their highly sensitive bills—and feed on materials such as silt-covered seeds, insect larvae, and relatively slow-moving and often bottom-dwelling prey that can be obtained without rapid chases or frequent failures in capture.

Durations of diving during foraging probably vary greatly with ecological conditions, such as water depth and food abundance, and so interspecific comparisons must be made with caution. However, it is possible to make comparisons when the water and food conditions are uniform, as at the Wildfowl Trust. Table 6 summarizes independent observations by Carbonell and Johnsgard on stifftail foraging-diving behavior in captive adults of three *Oxyura* species and the white-backed duck in the relatively shallow (usually under a meter deep) ponds of the Wildfowl Trust, plus a few observations by Johnsgard on a single wild adult male musk duck foraging in water of unknown depth and some observations by Weller (1968a) on wild black-headed ducks. At least under captive conditions, this species only rarely dives while foraging, although tipping up may often occur. Foraging dives have been observed among juveniles and adults in the wild, but in water less than a meter deep.

Recent observations by Hughes (1992) of wild ruddy ducks foraging in waters 1.5 to 5 meters deep revealed substantially longer average diving times than those shown in Table 6, perhaps because of the greater water depths involved. He found that the daytime dive durations of males in nonbreeding (basic) plumage and those of females and unsexed juveniles were not significantly different, but that

TABLE 6
Dive Durations (in Seconds) of Stifftails and the White-backed Duck

	Males*			Females			Both Sexes		
	N	Mean	S.D.	N	Mean	S.D.	N	Mean	Range
Black-headed duck†									
Dives		—			—		76	11.4	3–14
Pauses		—			—		29	7.0	2–12
Dive/pause ratio		—			—			1.6	
Masked duck‡									
Dives		—			—		16	21.0	11–26
Pauses		—			—		12	11.5	8–15
Dive/pause ratio		—			—			1.8	
North American ruddy duck§									
Dives	172	27.2	0.6	184	26.6	0.6			—
Pauses	172	10.9	0.3	184	11.1	0.3			—
Dive/pause ratio		2.4			2.4			—	
Maccoa duck	32	14.2	6.4	32	13.7	5.1	64	13.95	—
Argentine blue-billed duck*		—			—		8	24	21–26
White-headed duck	35	14.2	4.5	35	16.6	4.6	70	15.4	
Musk duck									
Dives	14	24.4	(range 15–32)		—			—	
Pauses	13	15.5	(range 10–25)		—			—	
Dive/pause ratio		1.6			—			—	
White-backed duck									
Dives		—			—		14	20.5	13–30
Pauses		—			—		13	14.5	9–28
Dive/pause ratio		—			—			1.4	

*Data for males are birds in nonbreeding (basic) plumage.
†Observations of Weller (1968a).
‡Observations of Jenni (1969).
§Observations of Hughes (1992); diurnal observations during winter in water 1 to 1.5 meters deep.

the dives of males in breeding (alternate) plumage were significantly longer (30.5 seconds) than those the other two groups. Although his nocturnal data were more limited, Hughes's studies produced very similar results, with males again having slightly longer mean dive durations than females. Interestingly, foraging activities occupied only about 7 percent of total daylight hours for ruddies but more than 60 percent of their nighttime activities, suggesting that about 90 percent of all foraging by ruddy ducks may be done at night.

The diving "pauses" shown in Table 6 are interdive intervals. These are usually assumed to represent minimum intervals required for resting between successive dives. Thus they have often been used as rough estimates of diving efficiency, reflecting the probable degree of exertion used during diving. The varied sources of these data, and their widely differing conditions, make direct comparisons impossible. However, the musk duck would appear to be an efficient diver

TABLE 7
Mean Dive Durations (in Seconds) of Stifftail Ducklings

| | Age of Ducklings, Weeks* | | | |
	1	2	3	3+
North American ruddy duck	10.6 (49)	11.1 (23)	13.9 (19)	15.3 (7)
Maccoa duck	11.0 (152)	11.3 (40)	12.1 (40)	14.1 (40)
White-headed duck	8.7 (75)	8.9 (20)	11.5 (40)	10.0 (40)

*Sample size in parentheses.
Source: Data of Carbonell (1983) from the Wildfowl Trust.

among these species, both as to its mean dive durations and its relatively short resting intervals. The black-headed duck may be the least efficient diver, judging from its short mean dive duration, which is in agreement with the relative anatomical specializations for diving exhibited in this species. In a detailed study of dive-to-pause durations among wild ruddy ducks in England, Hughes (1992) analyzed interdive intervals relative to dive durations and found a gradual increase in mean interdive intervals, with increasing mean dive durations up to dives lasting about 30 seconds, followed by a leveling off of the interval lengths with longer dives. This relative dive-to-pause duration relationship was considerably lower in its mean rate of incline (namely, the inverse of the dive-pause ratio, or 0.29 for females and 0.35 for males) than is the case with pochards (0.64) or mergansers (0.49) collectively as reported by Ydenberg (1986). Instead, it was was similar to the average slope rate (0.31) collectively calculated by Ydenberg for various loons, grebes, cormorants and alcids. As suggested by Ydenberg, this slope may not only reflect diving efficiency but may also be influenced by varied dietary preferences and differing rates of recovery from diving fatigue.

Table 7 summarizes dive-duration data for three species of *Oxyura* ducklings obtained by Carbonell at the Wildfowl Trust over several weekly periods of early duckling growth. What is especially surprising is the rapid rate at which stifftail ducklings begin to dive for durations that are almost as long as those typical of adults of these species. Joyner (1975) provided similar data for unfledged wild North American ruddy ducks, indicating that those up to a week old have dive durations averaging 6.4 seconds, those in their second week 8.2 seconds, in their third week 10.6 seconds, and in their fourth to seventh weeks 11.7 seconds. These durations were generally longer than those of adults feeding in water of comparable depth, as shown in Table 6. Siegfried (1973d) similarly estimated 5-day old and 15-day-old ruddy duck ducklings to have a mean dive duration of 12.9 seconds, whereas the dives of 25-day-old ducklings averaged 14.8 seconds.

Some information on adult foods of representative stifftails, plus the freckled duck, are summarized in Table 8. Unfortunately, nearly all of these analyses were based on gizzard contents, which greatly overestimates the amount of hard foods such as seeds that are ingested and correspondingly underestimates the amount of soft foods such as soft-bodied insects. More detailed descriptions of these species' foods will be found under the individual species accounts, but some general dietary patterns might be mentioned here. The freckled duck exhibits a diet very high in vegetable matter, especially soft, leafy materials, and relatively low in animal materials (aquatic insects, sponges, and very small crustaceans). The limited information on the foods of the black-headed duck suggests a similar high-vegetable, low-animal diet. The few species of *Oxyura* for which rather good dietary information is available seemingly have diets composed of both plant and animal materials, but the animal materials are composed almost enirely of midge (Chironomidae, Diptera) larvae obtained from the muddy bottoms of ponds and marshes. Finally, in the musk duck there is a predominance of animal foods in the diet, which is very low in midges but high in larger aquatic insects, especially aquatic bugs and larval Odonata (dragonflies and damselflies).

BREEDING AND NONBREEDING HABITATS

The least typical stifftail, the black-headed duck, is best adapted to living in dense marsh vegetation having small open pools, where it forages on seeds of marsh plants and sometimes on duckweed and snails. This habitat places the black-headed duck in close contact with coots, various ducks, and the numerous other marsh-dwelling species upon which it depends for its social parasitism (Weller 1968a).

All the other typical stifftails are also essentially marsh-dwelling, or at least marsh-breeding birds, dependent upon rather heavy growths of reedbeds for their nests, and upon adjacent areas of open water of varying depths and muddy bottoms rich in organic materials and aquatic insect larvae. During the nonbreeding season some species such as the Australian blue-billed duck may gather on fairly large, clearwater lakes and spend all of their time in open water far from shore, at least during daylight hours. However, as night approaches they move closer to shore and begin foraging during late afternoon, with some (perhaps most) feeding activity extending through the night and into early morning hours. Similarly, nonbreeding musk ducks are often seen on large lakes, far from shore, where they may feed in water at least 20 feet (ca. 6 meters) deep (Frith 1967).

The Argentine blue-billed duck favors large lowland lakes and semiopen marshes with large pools, as well as open roadside marshes that are connected by waterways to larger marshes (Weller 1967b). However, the Peruvian ruddy duck is found is rush-lined and reedbed-edged lakes at altitudes from as low as about 15 meters in Tierra del Fuego to as high as about 4400 meters in northern Chile (John-

TABLE 8

Reported Diets of Freckled Duck and Representative Stifftails

		Foods Present	
	Sample Size	Frequency,* %	Volume,† %
Freckled duck			
Frith, Braithwaite, and McKean (1969)	119 (adults)		
Ceratophyllaceae		—	23.8
Characeae		—	20.0
Cyperaceae (seeds)		—	12.3
Other plant foods		—	37.8
Total plant foods		—	93.9
Aquatic insects		—	3.0
Other animal foods		—	3.1
Total animal foods		—	6.1
Black-headed duck			
Weller (1968a)	27 (full-grown)		
Scirpus (seeds)		89	—
Gramineae (seeds)		3	—
Lemnaceae (all)		Present	
Mollusca (snails)		18	—
North American ruddy duck			
Cottam (1939)	163 (adults)		
Potamogetonaceae		—	30
Scirpus (seeds)		—	16
Other plant foods		—	26
Total plant foods		—	72
Chironomidae (larvae)		—	22
Other animal foods		—	6
Total animal foods		—	28
Australian blue-billed duck			
Frith, Braithwaite, and McKean (1969)	130 (adults)		
Typha (seeds)		74.7	8.5
Ceratophyllum		43.7	8.2
Myriophyllum (seeds)		43.5	6.7
Other plant foods		—	29.0
Total plant foods		100	52.4
Chironomidae (larvae)		78.8	26.7
Other animal foods		—	20.9
Total animal foods		100	47.6
Musk duck‡			
Frith, Braithwaite, and McKean (1969)	399		
Typha (seeds & leaves)		—	7.1
Potamogetonaceae		—	5.9
Other plant foods		—	9.8
Total plant foods		—	22.8
Hemiptera		—	24.6
Odonata (larvae)		—	10.3
Other insects		—	8.6
Mollusca		—	16.2
Crustacea		—	10.8
Other animal foods		—	2.1
Total animal foods		—	72.6

*Frequency of occurrence of food category as percentage of total specimens sampled.
†Volume of foods composing this food category relative to total volume of foods present in all specimens.
‡The musk duck sample included 62 ducklings, 189 immatures, and 148 adults. Adults had lower mean volumetric averages for total animal foods (66 to 69%) than did either immatures (92.7%) or ducklings (85.2%) and had correspondingly higher averages for plant foods. Higher ratios of animal foods among ducklings and immatures than among adults are also typical of Australian blue-billed ducks (Frith, Braithwaite, and McKean 1969) and North American ruddy ducks (Cottam 1939) and are generally typical of waterfowl collectively.

son 1965). North American ruddy ducks use both large and small marshes for nesting. Nests are located in emergent vegetation types, depending of the particular marsh or pothole being used, so long as the vegetation is growing at a depth (ca. 0.25 meter) most desirable for nesting (Bellrose 1980). The white-headed duck needs ample tracts of open water flanked by dense stands of emergent aquatic plants for cover and nest sites, as well as pathways for underwater escape (Cramp and Simmons 1977). Maccoa ducks generally favor ponds with stretches of open water near areas of tall emergent vegetation (*Phragmites, Scirpus,* etc.); these ponds typically have areas relatively free of floating or partially submerged aquatic vegetation (Macnae 1959). The Australian blue-billed duck prefers deep water in large, permanent freshwater wetlands, where ecological conditions are stable and aquatic plants are abundant; when breeding, this species seeks rather deep wetland areas that are rather densely vegetated throughout or that have extensive margins of emergent rush and sedge vegetation. The musk duck has very similar requirements. It prefers deep waters of large, permanent freshwater swamps where dense marginal vegetation provides breeding sites similar to those of blue-billed ducks (Marchant and Higgins 1990).

DISTRIBUTIONS, DISPERSALS, AND MIGRATIONS

To a considerable degree the stifftails are associated with the warmer parts of the world, virtually all species area being restricted to—or at least most common in—temperate to subtropical land areas situated between the tropics of Cancer and Capricorn. Only one species, the North American ruddy duck, breeds beyond 60 degrees of north latitude; in Eurasia the white-headed duck probably rarely breeds north of 45 degrees north latitude. The Argentine blue-billed duck's breeding range probably reaches its limit at about 50 degrees of south latitude in the Southern Hemisphere, as perhaps does that of the Peruvian ruddy duck. However, all stifftail species have at least some populations breeding in the tropical to temperate zone occurring between 30 degrees of north and south latitude. This generally warm-weather distribution probably limits the distances of migration required for these relatively short-winged birds.

Excepting the aberrant black-headed duck, all stifftail species have relatively small, short wings and take flight only with considerable difficulty. If not wholly sedentary, stifftails typically do not range far or migrate great distances, and so daily and seasonal food needs are perhaps thereby reduced. Thus the wing loading of a ruddy duck weighing 535 grams was estimated by Poole (1938) as 1.6 grams per square centimeter of total wing area, as compared with much lighter wing loadings of 0.89 grams for a wood duck (*Aix sponsa*) and 1.01 for a gadwall (*Anas strepera*) of about the same body weight. These figures would suggest that about 55 to 60 percent more energy must be expended in taking off and flying by ruddy ducks than by gadwalls or wood ducks. Humphrey and Livezey (1982) also estimated the mean wing loading of 13 North American ruddy ducks (averaging 530 grams) at 1.6 grams per square centimeter and that of 7 masked ducks (averaging 363 grams) at 1.03, the latter about the same as they estimated for the similar-sized (393 grams) bufflehead (*Bucephala albeola*)—at 1.16 grams. Briggs (1988) calculated the buoyancy index (a more reliable estimate of wing loading when comparing birds of greatly differing weights) as slightly greater (ie., lower wing loading) than that of the Australian blue-billed duck, which presumably represents an advantage for sustained flight for the North American species, a much more highly migratory bird than the Australian one. Briggs also judged that females of this latter species, and probably also of the relatively more sedentary musk duck and maccoa ducks, can afford to put on more prebreeding energy reserves than can wintering North American ruddy ducks, which must fly considerable distances before breeding.

Even more extreme, the wing loading of musk ducks has recently been estimated at more than 2.5 grams per square centimeter, which suggests that males weighing more than 2400 grams and females exceeding 1635 grams may be effectively flightless (Marchant and Higgins 1990). Musk ducks fly only rather rarely—and then seemingly get off the water only with great difficulty. During several weeks of observation Johnsgard never saw wild adult males in flight, and females were seen taking flight only a few times. On those occasions the birds took off from a lake surface only after a long running start into a strong headwind, cleared the water by less than a meter, and then crash landed after covering only about 30 to 40 meters in actual flight (Johnsgard 1965b).

The black-headed duck lies at the other stifftail extreme as to probable wing loading (no actual estimates are yet available), and it resembles the dabbling ducks in that the birds are able to rise quickly from the water. They sometimes forcibly strike the water with their wings during takeoff as do many dabbling ducks, and once airborne they can fly quite rapidly, although flight speeds have not yet been estimated (Weller 1968a).

In comparison with the true stifftails, freckled ducks are able to take off with a short running start and once airborne can fly with considerable speed and agility, with wing-beat rates of about eight per second, comparable to that of mallards (Johnsgard 1965b). No wing-loading estimates or flight-speed data are yet available for this species. In contrast, white-backed ducks take flight only rarely—and then attain flight only after a running start of 3 to 7 meters along the water surface (Brown, Urban, and Newman 1982).

In addition to some regular if short seasonal migrations, nomadic dispersal tendencies and opportunistic invasions of temporary wetland habitats appear to be typical for both musk ducks and Australian blue-billed ducks. According to Frith (1967), the musk duck sometimes undertakes considerable movement to exploit newly formed habitats, such as periodically colonizing Lake Eyre in south-central Australia whenever it contains water. Yet musk ducks are only rarely seen in flight, and their relatively small sternal muscles would suggest that long-distance migration must be made as a series of short hops. This limited flying potential may help explain why the birds sometimes appear temporarily in isolated tanks on the plains, as reported by Frith.

Although the Argentine blue-billed duck must normally have only quite short migrations, Todd (1979) reported that during the summer of 1916–17, following a severe drought in Argentina, a number of these birds appeared on Deception Island, some 1000 kilometers offshore, indicating a considerable potential for long-distance dispersal.

The North American ruddy duck is probably the most strongly migratory of all the stifftails. Some of its northernmost populations may regularly migrate as far as 1500 kilometers, although few specifics are available. At least one ruddy duck banded in southern Saskatchewan was subsequently recovered about 2600 kilometers away in Chihuahua, Mexico (Smith 1949). Generally, ruddy ducks from interior breeding areas of North America migrate to the Atlantic or the Pacific Coast. These latter birds then sometimes move southward to the western coast of Mexico. Some birds also fly from the Great Plains south to the Gulf Coast, and others move eastwardly to New England and Chesapeake Bay. Probably much of the migration occurs at night, the birds moving in relatively small flocks (Bellrose 1980).

Ruddy duck stragglers have reportedly reached the Hawaiian Islands on three occasions (Hawaii twice, Oahu once). These islands are about 3500 kilometers west of the Pacific Coast and probably represent the extreme limits of mobility for the species. In England, the introduced North American ruddy duck population makes short seasonal movements of up to 70 kilometers (Cramp and Simmons 1977). They have also crossed the English Channel and become self-introduced in various countries of western Europe, where they have encountered and become locally sympatric with the native white-headed duck in Spain. They have also been reported from Iceland, which is almost 1000 kilometers north of Scotland, and there is a recent report of an occurrence in the Ukraine.

COMPETITORS, ECOLOGICAL ASSOCIATES, AND NESTING HOSTS

By and large, a maximum of only one species per genus of stifftail occurs in most areas, so strong interspecific competition is seemingly reduced. Three species of *Oxyura* as well as the monotypic *Heteronetta* occur in South America—the group may therefore be presumed to have originated there—and only in South America does any significant opportunity exist for ecological interactions among closely related forms. In Chile, the Peruvian ruddy duck and Argentine blue-billed duck have overlapping ranges and limited sympatry (Casos 1992), but the former is primarily an Andean form and the latter occupies lowland lakes (Johnson 1965). In Argentina the Argentine blue-billed duck is locally sympatric with both the masked duck and the black-headed duck. Surprisingly, however, neither the masked duck nor the blue-billed duck is yet known to have been parasitized by the black-headed duck (Weller 1968a).

Probably the most significant interspecific vertebrate interactions of any of the stifftails involve the black-headed duck, which is evidently wholly dependent upon other species for incubating and hatching its eggs. Weller (1968a) listed at least 12 species of birds that have been reported in the literature as host species for the black-headed duck; he found two additional host species during his own fieldwork. Nearly all of these were anseriform, charadriiform, ciconiiform, and gruiform birds that nest in dense marsh vegetation. The highest parasitism incidence observed by Weller involved red-fronted coot (*Fulica rufifrons*), red-gartered coot (*F. armillata*), rosy-billed pochard (*Netta peposaca*), and white-faced ibis (*Plegadis falcinellus*).

In the case of the red-fronted coot, 87 percent of that species' unparasitized nests and 81 percent of its parasitized nests were judged successful. These host nesting success rates (defined as the percentage of nests hatching at least one host chick) suggest that any reduction in nesting success caused by parasitism by the black-headed duck is probably only on the order of about 10 percent. The ranges of these two species closely coincide, and their habitat preferences are very similar. These facts caused Weller to speculate that the red-fronted coot is the species most likely to have played a key role in the evolution of parasitic behavior in black-headed ducks, since the coots are abundant, are tolerant of foreign eggs, show broody behavior toward newly hatched ducklings, and have a high rate of nesting success. Additional details on these and other interactions between the black-headed duck and its varied host species may be found in the account of this species in Chap. 6.

Females of many, perhaps all, species of *Oxyura* frequently lay eggs in the nests of other ducks, often including those of their own species as well as in nests of various other marsh-nesting duck species. For example, Joyner (1975) found that 62 of 809 duck nests studied in Utah were parasitized interspecifically by North America ruddy ducks, with those of cinnamon teal (*Anas cyanoptera*), northern pintail (*A. acuta*), common mallard (*A. platyrhynchos*), and redhead (*Aythya americana*) all being affected. Redheads were the species most often parasitized (12 percent of 93 nests), followed by mallards (7 percent of 29 nests), northern pintail (6 percent of 48 nests), and cinnamon teal (2 percent of 474 nests). The mean number of ruddy duck eggs per host nest varied greatly, from 5.5 per nest in mallard nests to 4.0 in pintail nests, 2.5 in cinnamon teal nests, and 1.5 in red-head nests. Host hatching success (defined as the percentage of host eggs in nests that hatched successfully) in unparasitized mallard nests was 65.1 percent of 130 eggs, as compared with 18.2 percent for mallard eggs in 2 nests parasitized by ruddy ducks. The hatching success of unparasitized pintail nests was 62 percent of 138 eggs, as compared with 28.6 percent of pintail eggs in 3 nests parasitized by ruddy ducks. The hatching success of 1645 cinnamon teal eggs in unparasitized nests was 58 percent, which was comparable with the 62 percent success rate determined for 11 parasitized nests. In redheads the corresponding hatching success percentages were 36.8 of 825 eggs in unparasitized nests versus 2.8 of redhead eggs in 10 parasitized nests.

The overall total estimated hatching success of all parasitically laid ruddy duck eggs in Joyner's study was 17.6 percent for 68 eggs, as compared with an estimated "control" hatching success rate of 31.2 percent for 1419 ruddy duck eggs in self-incubated ruddy duck nests that were themselves unparasitized by redheads. Joyner judged this success rate to be perhaps lower than normal for the species generally, owing to local flooding effects in his study areas and higher reported hatching success rates by other observers—e.g., estimates of 52 percent by Williams and Marshall (1938) and 69 percent by Low (1941). The average three-year estimated nesting success of self-incubated and unparasitized ruddy duck nests was 34 percent for 152 nests, with a mean of 8.5 young hatched per successful nest. Under these circumstances the average annual production of hatched young per female incubating its own nest may be estimated to be 2.9 young. Given an estimated hatching success rate of 17.6 percent for parasitically laid eggs, in order for a female ruddy duck to produce more young by laying eggs parasitically than by incubating her own eggs she would have to deposit at least 17 eggs in other duck nests. This figure is about twice the average clutch size of ruddy duck nests found by Joyner in his study area (9.3 eggs for 152 nests) and would seem to be a basically inefficient if not physiologically impossible mode of obligatory reproduction for ruddy ducks. Additionally, Joyner found that ruddy duck females can successfully hatch nests containing up to as many as 16 eggs with a hatching success rate in excess of 30 percent, so there would be no apparent reproductive benefit in using other birds' nests to deposit eggs that would increase clutch size beyond some specific optimum number. However, to the extent that a female might be able not only to produce and incubate its own clutch but also be able to deposit a few additional eggs in other birds' nests (perhaps including those of other ruddy ducks), such behavior could increase her reproductive output slightly without perhaps imposing too much physiological stress. In Joyner's total sample of 1449 nonparasitically laid ruddy duck eggs, approximately half of these were found in clutches (including probable dump-nests) of 11 eggs or more, suggesting that perhaps ruddy ducks are indeed physiologically able to lay more eggs than their seemingly optimum clutch size (in terms of mean hatching success) of 7 to 9 eggs. Indeed, captive female ruddy ducks have been found to be able to lay up to four clutches in a single season (Murton and Kear 1978).

Gray (1980) estimated that about 1260 kilocalories would be needed to produce an average North American ruddy duck clutch of seven eggs, but much of that requirement would already be provided by the 1350 kilocalories typically previously stored as reserve body lipids by females. Assuming a 77 percent efficiency in converting and transporting body reserves into eggs, this conversion cost was 294 kilocalories, spread out over the prelaying and laying periods. Assuming that 5000 chironomids per hour might be consumed in good foraging areas, these collectively represent about 30 kilocalories of food intake per hour. About 7 to 10 hours per day are spent by prenesting females in foraging, easily providing the estimated 217-to 273-kilocalories daily energy intake that might be required. Carbonell (1983) independently but similarly calculated that a clutch of six North American ruddy duck eggs would collectively cost the female 1202 kilocalories and that a daily energy cost during the egg-laying period of 199 kilocalories would be incurred by the female. Gray estimated that the collective energy costs of reproduction were 79 percent higher for females than for males and that feeding requirements for females (number of prey items consumed) were 67 percent greater than for males.

Probably little direct competition for nest sites occurs among stiff-tails and other similarly water-dependent nesting ducks. Consider the interactions of nesting ruddy ducks and canvasbacks (*Aythya valisineria*) and redheads in North America. Featherstone (1975) observed a substantial amount of niche overlap in nest-site selection criteria of these three species in southern Manitoba, the common criteria involved such parameters as the density of vegetation at the nest site, water depth at the nest, distance from the nest to nearest open water, and pond size. He judged that interspecific niche differentiation was mainly achieved by asynchronous nesting in these three species—and secondarily achieved by variations in vegetation density at the nest.

Less easily measured food- and foraging-related competitive interactions probably occur regularly between stiff-tailed ducks and other diving ducks. Thus Amat (1984) reported that his estimates of niche breadth and niche overlap between the white-headed duck and two sympatric species of pochards—the Eurasian pochard (*Aythya ferina*) and the tufted duck (*A. fuligula*)—indicated a degree of niche complementarity. Of the three, the pochard showed the greatest flexibility in its feeding behavior, the tufted duck the greatest flexibility as to horizontal distances progressed underwater, and the white-headed duck the greatest flexibility in feeding zones utilized. The greater the amount of overlap in feeding zones used by the two *Aythya* species, the greater the observed utilization of these same zones by the white-headed duck, suggesting that foraging competition might indeed exist. Siegfried (1976C) did a similar study in Manitoba with the North American ruddy duck and three other diving ducks of the genus *Aythya*—namely the canvasback, redhead, and lesser scaup *(A. affinis)*. He concluded that selection for different foraging sites within the same general wetland habitat type (prairie marshes) may have been more important in achieving foraging niche segregation among these four species than was selection for different food types.

In Woodin's (1987) studies of breeding-season interactions between redheads and North American ruddy ducks in North Dakota, he found that ruddy ducks were more carnivorous, their prebreeding and breeding-period foods averaging about 90 percent invertebrates (mostly chironomids). By contrast, redheads were more omnivorous, consuming from 30 to 49 percent plant materials, although chironomids again were the most important single food item during the prelaying and laying periods. Food selection in both species was related to reproductive status and also, in some cases, to relative food abundance. Wetland habitat selection in both species was affected by varied water regimes and wetland size classes. As compared with about 12 other duck species breeding in interior North America, female ruddy ducks during the laying phase consume the highest average proportion of animal matter, especially insects, and the lowest proportion of plant matter (Krapu and Reinecke 1992).

Argentine blue-billed duck, pair swimming.

CHAPTER FOUR

Comparative Social and Sexual Behavior

Pair bonding for a single breeding season is believed to be the typical pattern of most waterfowl in the subfamily Anatinae (tribes Cairinini, Anatini, Aythyini, Mergini, and Oxyurini), with the majority of the displays being performed by the male as birds of that sex compete for the generally less abundant adult females in the population. Kear (1970) tabulated the associated pair-bond correlates in these groups, such as sexual dimorphism tendencies in plumages (males brighter, at least seasonally), the variably larger size and more aggressive tendencies in males, and the age at sexual maturity, which sometimes occurred even before the definitive adult breeding plumage has been attained.

ADULT SEX RATIOS AND THEIR BIOLOGICAL IMPLICATIONS

With the unbalanced sex ratios favoring a predominance of males in nearly all adult duck populations, the degree of competition for mates is correspondingly increased. Thus males must spend more time displaying both sexually and aggressively, and selection pressures for aggressive male-male interactions are perhaps at least as intense and important in achieving sexual success as is selection associated with heterosexual attraction. Females must eventually choose from among the available males, at minimum for achieving egg fertilization and perhaps additionally for preventing other males from constantly harassing them sexually while they are attempting to incubate eggs or lead dependent young. Pair bonds are certainly substantially weaker in stifftails than in most ducks. This behavior possibly reflects the fact that female stifftails are unusually adept at evading males by diving or by threatening them, and the newly hatched young of stifftails are highly precocial and have remarkably limited needs for maternal care. These factors have allowed females a certain degree of emancipation from dependency on males—and similarly have emancipated males from remaining with females for longer periods than required to assure their fertilization.

The basic biological reason for unbalanced sex ratios in waterfowl is believed to result primarily from the greater susceptibility of females to predation while on the nest—as well as their possibly greater vulnerability to starvation or freezing because of their smaller size and the periodic stresses associated with egg laying and incubation. Female ducks may also be more vulnerable to hunters than are males, at least among those species whose males are not specifically shot for their trophy value (Hochbaum 1939). For example, among Wisconsin hunter-killed ruddy ducks, females composed 54 percent of the total harvested sample of 385 birds (Jahn and Hunt 1964). However, spring sex-ratio counts for all major areas of the species' range exhibit a predominance of males in the population. Such counts collectively indicate that males represent from about 59 percent (Johnsgard and Buss, 1956) to 61 percent (Bellrose et al. 1961) of spring North American ruddy duck populations.

As to the other stifftail species, Johnsgard (1965b) similarly reported that males composed 62.6 percent of 711 Australian blue-billed ducks that he observed during late spring on a breeding-ground lake in Victoria. Weller (1968a) likewise reported that males composed 58 percent of a total of 548 adult black-headed ducks that he observed in Argentina. On the other hand, Amat and Sanchez (1982) tabulated sex-ratio data for 272 white-headed ducks in Spain, of which only 46.7 percent were identified as males. This result represents an apparent exception to the male-predominance situation among waterfowl generally, but possibly some femalelike first-year males were confused with females in this sample. In general, it would seem that there are from about 1.4 to 1.6 adult males per female in courting and breeding stifftail populations, which means that about a third of the available males are unlikely to obtain female mates, assuming a monogamous mating system. This divergent sex ratio in itself provides a strong basis for sexual selection, and if some males are nonmonogamous and the more dominant ones are able to fertilize more than one female, leaving even fewer for the others to inseminate, the potential sexual selection pressures are obviously even greater.

Regrettably, no good sex-ratio data are yet available for the musk duck. In spite of its large size and conspicuousness, it is difficult to obtain good sex-ratio data on this species in the field. In Johnsgard's experience it seems to show considerable spatial segregation of the sexes during the nonbreeding season, when some limited assemblages of females and immatures do develop. Furthermore, in breeding areas the birds are well dispersed, with the females becoming increasingly secretive and the adult males progressively more conspicuous. Consequently, the two sexes are very difficult to survey accurately during either season. Among specimens that apparently were collected and measured by Frith (1967) there was a distinct predominance of females (292 females, 243 males), which probably simply reflects the apparently greater vulnerability of females to hunters, as was mentioned earlier for the North American ruddy duck.

In correlation with the unbalanced sex ratios of stifftails, their varied tendencies for polygynous matings, and the associated variable potentials for sexual selection, there might also be variably unbalanced adult body mass ratios in these birds. That is, species having the most polygynous mating systems and divergent adult sex ratios should also have the largest and heaviest males relative to females, since the larger and stronger males should be able to socially dominate the smaller and weaker individuals. From the adult weight data summarized in the species accounts of Chaps. 6 through 13, the estimated male-to-female mean adult mass ratios for various stifftails (organized by increasing relative male mass) are as follows: blackheaded duck 0.9:1; masked duck 1.0–1.1:1; Australian blue-billed duck 1.1:1; North American ruddy duck 1.05–1.1:1; white-headed duck 1.04–1.1:1; Argentine blue-billed duck 1.2:1; maccoa duck 1.12–1.3:1; and musk duck 1.45–1.54:1. This increasing divergence in adult weights roughly parallels known relative pair-bonding and aggression tendencies in the stifftails, with the black-headed duck having the strongest pair bonds and the males being relatively small and less aggressive. At the opposite extreme, the musk duck has no apparent pair bonds and the large males are by far the most aggressive. The genus *Oxyura* connects these extreme types. Three *Oxyura* species that may have relatively monogamous or only weakly polygynous pair bonds (masked duck, Australian blue-billed duck, and North American ruddy duck) have adult mass ratios of 1.0–1.1:1, whereas the distinctly polygynous white-headed duck and maccoa duck have ratios of 1.1–1.3:1. Weight data on the possibly monogamous Argentine blue-billed duck are still too limited to fit confidently into this sequence but would seem to fall into the latter (presumptively polygynous) category in terms of sexual mass ratios.

The slightly reversed sexual dimorphism in body size of *Heteronetta* is worth noting, as it is perhaps unique in the entire family Anatidae. In this case it can presumably be attributed to selection for larger body size in females so as to facilitate the laying of more eggs, in conjunction with its obligate social parasitism. If that argument can be applied to *Oxyura* too, then the masked duck and Australian bluebilled duck should hypothetically either have unusually large clutch sizes or perhaps be at least semiparasitic, since the females are about as large as the males. Yet, based on available data, neither of these species has especially large clutch sizes. Carbonell (1983) estimated the average energy investment of female stifftails, stated as total average clutch weight relative to average adult female weight, as 80 to 90 percent in North American ruddy (assuming a 6-egg clutch), maccoa (5.9-egg clutch), and white-headed ducks (5.8-egg clutch); 60 percent in the Argentine blue-billed duck (4.3-egg clutch); 54 percent for black-headed duck; and 12 percent for musk duck (2-egg clutch). If slightly differing weight assumptions are used, these ratios work out as 90 percent for the masked duck; 80 to 90 percent for the maccoa, North American ruddy, and white-headed ducks; 62 to 64 percent for the black-headed duck and the two blue-billed ducks; and 21 percent for the musk duck. (See Table 16). Perhaps the relatively large body sizes of possibly monogamous stifftail species such as black-headed, masked, and blue-billed ducks are related both to maximizing their egg production abilities and to "equalizing" their social interactions with males.

INTERSPECIFIC VARIATIONS IN PAIR BONDING

Obviously, seasonal intensity of display in any duck species will probably reflect the local climatic and especially the photoperiodic conditions of the location. Thus interspecific comparability is difficult unless two or more species can be studied concurrently in the same area. This is one of the advantages provided by collections of breeding captive birds such as that of the Wildfowl Trust, where Carbonell (1983) observed the breeding behavior of six species of stifftails over a three-year period. However, a corresponding disadvantage of studying captive birds is the possibility that quantitative or qualitative aspects of behaviors performed under these conditions may differ from those in the wild, leading to erroneous conclusions.

Besides an apparent total lack of pair bonding and the presence of a promiscuous mating system in the musk duck (Johnsgard 1966), one finds variable degrees of weakened monogamy and its replacement by male-dominance polygyny or promiscuity in other stifftails.

A rather weak but sometimes permanent pair bonding seems to be present in the black-headed duck (Weller 1968a). At the Wildfowl Trust, one female was paired to a male for only several weeks, whereas another that lost her mate during the breeding season did not pair again during that year. However, a good deal of apparent "raping" or extra-pair copulatory behavior was seen in these birds. During these activities unpaired females would be chased by other paired or unpaired males, the paired males leaving their own mates to join in the chase. Paired females were sometimes also sexually assaulted, although their mates almost always attempted to chase the intruding male(s) away (Carbonell 1983).

Rather weak pair bonds were formed in North American ruddy

ducks at the Wildfowl Trust. Apparent pair bonding has also been observed in the wild (Joyner 1969, 1975), although some females have been observed being attended by more than one male (Gray 1980). The bond was apparently formed about a week before the first egg was due to be laid and lasted only until the clutch was completed. A bond was later reestablished only if the clutch was lost and the female was ready to lay again. Contrary to some other observers (e.g., Joyner 1975), no pair-maintaining behavior was observed during incubation. One dominant male was paired to two, sometimes even three, females simultaneously, and any male would typically display and copulate with females other than its mate so long as he was sexually active. However, apparent rape behavior was observed only twice, and in both cases was inexplicably directed toward a female New Zealand scaup (*Aythya novae-seelandiae*). Among wild ruddy ducks in California, Gray (1980) observed 25 cases of attempted conspecific rape during two years of study. Most of the females involved in the rape attempts were unmated. Only four attempts involving paired females were seen, and only one of these was successful. Eighty-three percent of the total rapes observed occurred during the egg-laying and incubation period, and 25 percent of these were judged successful. Some of these extra-pair copulations might at times be adaptive, inasmuch as renesting or second-nesting efforts might be undertaken.

No definite pair bonding was evident to Carbonell among the white-headed ducks at the Wildfowl Trust. This observation is contrary to those of some observers (Amat and Sanchez 1982) but in agreement with others (Matthews and Evans 1974; Veselovksy 1976). During two years, a single male had the attention and control of the four females present on a pond. In the third year this male was accepted by three females, while a second male was accepted by two females. Rape behavior was not observed.

Among the maccoa ducks at the Trust, no pair bonding was evident to Carbonell, a situation generally also believed to be the case under natural conditions (Siegfried and van der Merwe 1975). The females evidently chose a particular male with which to mate but did not necessarily nest within his territory. Males would remain with a particular female much of the critical breeding time, beginning from about a week before egg laying started and lasting until incubation began, but no apparent pair bonding occurred. Rape behavior was not observed. Siegfried (1985) has noted that in this species the presence of a dominant male can cause a suppression of the breeding plumage among other males on the same pond, which would probably reduce direct sexual and territorial competition among males.

Pair bonding was stronger than in other *Oxyura* species among the Argentine blue-billed ducks studed by Carbonell at the Trust. One pair remained together throughout the entire three years of study, but two other females did not remain paired after the breeding season.

A fourth female in the same pen was courted by two of the paired males as soon as their respective mates began to incubate. Raping behavior was common, and in 21 of 25 observed cases it involved the raping of females already paired to another male. In the other four cases it involved raping females of other species (rosy-billed pochards and black-headed ducks).

In the Australian blue-billed ducks pair bonds are evidently short-term, but pair-bonding strategies of this species are still uncertain. In one captive situation, one of three males in company with four females reportedly was able to dominate and control at least three of the females, whose clutches often overlapped temporally (Marchant and Higgins, 1990). From this limited information it would appear that this species may have a pair-bonding pattern less like that of the otherwise similar but relatively monogamous Argentine blue-billed duck and more like those of the other apparently polygynous *Oxyura* species.

Pair bonds are evidently completely lacking in musk ducks. This situation correlates well with the extreme sexual dimorphism of the adult sexes and the great amount of time and energy expended by males in performing their advertising displays. At the Wildfowl Trust the lone adult female was attacked and nearly killed by the dominant adult male before they were separated into adjacent pens, connected by a hole only large enough for the female to pass through. A second adult male was also present and shared the same pen area used by the single female. Unlike the other male, he was submissive to and constantly chased by this female. Although then a fairly old adult, this second male did not begin to display until both the dominant male and the female had died in 1980 and 1981, respectively. Copulation behavior was not observed by Carbonell at the Wildfowl Trust, and only a few eggs were ever laid there. However, rapelike copulations have been observed among wild birds, as have vicious, seemingly life-threatening fights between adult males (Johnsgard 1966).

INTERSPECIFIC VARIATIONS IN SEASONALITY OF SOCIAL DISPLAY

Generally speaking, male stifftails remain in breeding plumage and appearance over surprisingly long periods (typically about eight months) at the Wildfowl Trust. Carbonell found that these periods extended from about January or February until August or September, with wing and/or tail molting periods marking both the start and end of these periods (Fig. 10). All of the species were out of nuptial plumage and breeding condition from October through December. This situation suggests that Southern Hemisphere breeders normally breeding during those months (such as the Australian blue-billed duck) do shift their internal rhythms over to coincide entirely with Northern Hemisphere seasons, presumably under the influence

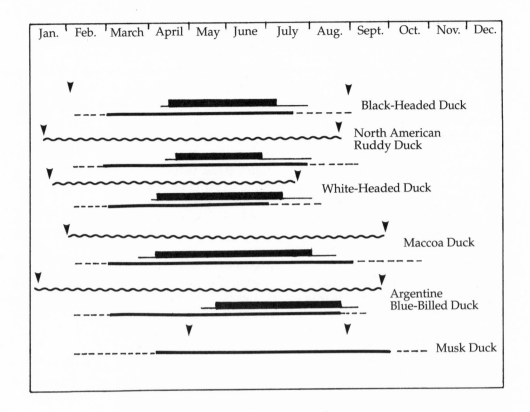

| Jan. | Feb. | March | April | May | June | July | Aug. | Sept. | Oct. | Nov. | Dec. |

Figure 10. Diagram of stifftail phenology (for 1980–82) at Wildfowl Trust, showing timing of wing and/or tail molts (pointers); durations of breeding plumage in males (wavy lines); egg-laying periods (bars indicate egg-laying periods occurring during all three years, extensions indicate extremes observed); and durations of male sexual displays (continuous heavy lines indicate regular display, broken extensions indicate sporadic display). (Redrawn from Carbonell 1983.)

and control of photoperiodic stimulation or perhaps seasonal changes in food availability.

According to Carbonell's observations, all the species of *Oxyura* as well as *Heteronetta* had relatively prolonged egg-laying periods that lasted from an average of as brief as about two and a half months (North American ruddy duck) to as long as four months (maccoa duck) and which, not surprisingly, were centered around the period of peak male display (Fig. 10). However, rather intense male display began well before egg laying started and typically persisted for a time beyond it. On average, the earlier species to begin laying were the more promiscuous ones—namely, the maccoa duck, white-headed duck, and North American ruddy duck. The relatively monogamous Argentine blue-billed duck was the latest to begin to lay, and male display in that species terminated almost immediately after the egg-laying period. Since the lone musk duck female only laid six eggs (three clutches of two eggs between early April and late May), no clear correlation between male display periods and egg-laying periods was possible for that species.

AGGRESSION, ALARM, AND ESCAPE BEHAVIOR

Stifftails at the Wildfowl Trust rarely chased other birds or one another, except during the breeding season. The least aggressive of all the species there was the black-headed duck, males of which would chase away other males who approached their females, typically by stretching the neck silently toward the intruder. If the intruder persisted he would be pecked, but actual fighting occurred only infrequently. However, females of *Oxyura* are notably aggressive during the breeding season, and when leading young, they have been observed to attack a variety of water birds much larger then themselves (Frith 1967; Joyner 1977a; Hughes 1992).

Among *Oxyura* species, aggressive encounters were frequent in courting groups. In both sexes the lowest-intensity aggressive posture is the "open-bill threat" (Matthews and Evans, 1974). It consists of the bird holding its head and neck forward, bill open and scapulars raised, and uttering a hissing sound. If the opponent does not move away, a more hostile posture is assumed, the "hunched rush" (Clark 1964; Johnsgard 1965a), in which the bird swims rapidly toward the intruder with its head and neck stretched forward and its scapulars raised. These postures are usually enough to drive the opponent away without direct fighting. However, fighting sometime does occur, the birds initially facing each other motionless with their tails held flat on the water and their necks stretched forward (pre-attack posture of Ladhams 1977). This phase lasts for several seconds. The birds then sometimes leap up in the air, very much like rails do when fighting, amid much splashing; the almost immediate result is the retreat of one of the two birds. Alternatively, one of birds might avoid direct confrontation and escape by diving or swimming away. Fighting is more frequent in the Argentine blue-billed duck than in the other *Oxyura* species. As with the maccoa duck, the blue-billed does not always face his opponent motionlessly prior to fighting. Rather intense fighting often begins when the aggressor catches up with the opponent; fighting occurs both on the surface and while the two are underwater.

Territorial expulsions of other birds by male maccoa ducks are performed in a "swimming-low-and-swift" posture (Clark 1964). In this posture the bird flattens his body, neck, and tail, with the bill touching the water, sometimes diving and later emerging while still in the same posture. Typical stiff-tailed ducks often precede an attack on another bird with a stealthy approach, the head lowered toward the water and the tail flat on the water, followed by a smooth, silent dive (here called a "stealthy dive").

In the musk duck, threats by males occur not as clearly ritualized postures but rather as a straightforward chase and attack, which often also involves diving. Both males also invariably respond to human approach by performing a splash-diving display, which has also been observed in wild birds (Lowe 1966), In this display the bird dips its head in the water while kicking its feet sideways and backwards, producing a large, loud splash as it disappears underwater, much like the noise produced when an alarmed beaver slaps its tail on the water while diving. This display seems to represent a combination of threat and escape behavior. Lowe also observed a somewhat similar display, the "travel-splash kick," in which the feet are kicked strongly, splashing water into the air while the head is kept in the normal swimming position. This posture was usually performed during a sequence of the typical "paddle-kick" advertising display and is of uncertain function beyond allowing the bird to move forward while also displaying.

Probably males of all *Oxyura* species perform the "ringing rush" (Johnsgard 1965a), although it has not been seen in the masked duck and is evidently rare in the white-headed duck. It has also been observed in some females, although it is much more characteristic of males. This posture, also called the "ring rush," "running flight" (Siegfried and van der Merwe 1975), and "display flight" (Gray 1980), consists of lowering the tail to the water, flattening the body and head feathers, stretching the neck forward in an alert manner, and then rushing along the water while furiously beating the wings and producing a distinctive popping or ringing sound with the wings or feet. (See Fig. 20.) This posture seems to have been derived from and sometimes still serves a general takeoff function. However, it also functions as a means of rapid and conspicuous semiaquatic locomotion, as may be required when changing location during social display and as a means of self-advertisement among males. When performed in this last-mentioned situation it is usually but not always oriented toward females. It occurs under various social circumstances, such as when a male is approaching a female who is being courted by other males, and less frequently as an apparent escape from other males after a bout of social display. It, or a similar nondisplay variant, also occurs in some nonsocial situations.

As noted earlier (Chap. 3), preflight movements are quite poorly developed and inconspicuous in stifftails. However, both Johnsgard and Carbonell have observed the North American ruddy duck performing head flicking or head shaking prior to performing the ring-rush display. Hughes has observed head shaking prior to takeoff in the ruddy duck. The maccoa duck has also been observed by Carbonell to perform head flicking and, though less often, to perform the swimming shake or to roll its cheeks on its back prior to starting the ring rush.

All stifftails are relatively quiet at most times, especially during alarm. When alarmed, *Oxyura* species at the Wildfowl Trust normally gather near the center of a pond, with their heads held high ("head-high" posture); they are either silent (North American ruddy duck and white-headed duck) or may utter soft purr-purr notes (Argentine blue-bill and maccoa ducks). Extended low-level alarm may lead to head pumping in both sexes, and this is occasionally followed by the "bubbling" display (see Fig. 20) in both sexes of the North American ruddy duck or by "sousing" in males of the maccoa duck and Argentine blue-billed duck (see Figs. 22 and 24). With sudden alarm the most common *Oxyura* reaction is to dive quickly and to assume a raised-neck, tail-lifted ("head-high, tail-cocked") posture after returning to the surface. When maccoa ducks are exposed to a sudden but prolonged alarm they quickly dive, emerge in the head-high, tail-cocked posture, and then quietly dive again without even making a ripple on the water.

Alarmed black-headed ducks also tend to swim to the center of a pond, but if the alarm occurs rather suddenly, males sometimes utter their courtship "toad call." (See Fig. 14.) A distinctive alarm or "wing-up, tail-up" posture, with strongly raised tail and folded wings (see Fig. 14F), was observed in wild birds by Weller (1968a) but has never been seen by Johnsgard or Carbonell among captives at the Wildfowl Trust. Wetmore (1926) additionally noted that disturbed wild black-headed ducks would quickly dive when startled, perhaps using their wings as well as their feet in submerging, and would then disappear in the manner of grebes.

SOCIAL DISPLAY CHARACTERISTICS

One of the basic aspects of pair-forming and pair-bonding behavior in nearly all ducks of the subfamily Anatinae is associated with female-inciting behavior (Johnsgard 1965a). In this behavior, females assume postures and use calls that identify more-favored drakes and, to varying degrees, "incite" them to threaten or attack other, less-favored ones. This behavior is relatively functional in groups such as shelducks (Tadornini). However, in more typical ducks such as dabbling ducks, pochards, and sea ducks, inciting behavior is highly ritualized, and male responses are much more likely to be sexually directed toward the inciting female than aggressively oriented toward the other male or males.

Female stiff-tailed ducks appear to lack a well-defined inciting behavior, as was first reported for *Oxyura* by Johnsgard (1965a) and later (1966) for *Biziura*. Weller (1968a) likewise observed no inciting behavior in his 11-month studies of *Heteronetta*. Although Gray (1980) used the term *inciting* in her descriptions of North American ruddy duck behavior, she admitted that this was apparently not a ritualized display in that species. Hughes (pers. com.) believes that these interactions in ruddy ducks do qualify as inciting, inasmuch as after a female has threatened or attacked her own partner this male may then threaten or attack other intruders.

Carbonell (1983) also failed to observe inciting behavior in any species of *Oxyura* or in *Heteronetta* during her three years of study. She provisionally attributed this absence of inciting to lack of pair bonding in some of these species (North American ruddy duck, maccoa duck, and white-headed duck) but also judged that, in the case of *Heteronetta*, inciting behavior may not be required if a permanent pair bond actually normally exists. She found no explanation for an absence of inciting in the apparently pair-bonding Argentine blue-billed duck, although she speculated that captive conditions existing at the time of her study may have influenced the pair-bonding tendencies she observed in this species. Nevertheless, regardless of their varied pair-bonding traits, inciting is seemingly lacking in all female stifftails.

No good theoretical explanation yet exists for the absence of inciting-based pair-bonding behavior in stifftails. Inciting is also apparently lacking in the seemingly primitive freckled duck, which evidently, like many stifftails, has only fragile pair bonds that persist for a relatively short period (Marchant and Higgins 1990). However, some continuing loose association of pairs may occur after breeding, and reformation of pairs in succeeding seasons may be common (Fullagar, Davey, and Rushton 1990). If indeed both of these groups diverged early on in anatid phylogeny, it is possible that this behavioral divergence occurred before inciting behavior had become entrenched as a basic pair-forming and pair-bonding mechanism in ancestral anatine ducks.

GENERAL ASPECTS OF MALE SOCIAL AND SEXUAL SIGNALS

Breeding male stiff-tailed ducks, especially those of the genus *Oxyura*, tend to be rather similar in their visual sexual signaling characteristics. These characteristics include (1) a bright blue bill, (2) an enlarged (inflatable) throat or neck, (3) a generally chestnut-brown body plumage, and (4) a black, or partly black, head color. These general collective traits approximate those of present-day *Heteronetta* males, which can be thus hypothesized as roughly resembling an ancestral *Oxyura* plumage type. Although the elongated tail of typical stifftails may also be important in display, the wings are not modified in color to provide potential social signals, except for the masked duck. Whether the male masked duck uses his white wing speculum during social display is still unknown; the patch might instead have other more general but still unknown social functions. Likewise, the feet of male stifftails are never colorful, in contrast to those of many male sea ducks. The feet are rarely apparent during sexual display (but are sometimes exposed during abdominal preening) and thus have evidently not been incorporated into visual social signaling.

Unlike dabbling ducks, most male stifftails are not highly vocal—indeed, most are relatively quiet—during social display, as are fe-

TABLE 9
Social Display Behavior and Comfort Movements

Movement	Ritualized Male Signal	Functional Comfort Behavior
Nibbling-preening	Dab-preening	Breast or foreneck preening
Oiling-preening	Cheek rubbing	Head rubbing (or shoulder rubbing)
Shaking or flapping	Swimming shake (or general shake)	Swimming shake
	Head shaking	Head shaking
	Head flicking	Head flicking
	Wing shaking	Wing shaking
	Wing flapping	Wing flapping
	Tail shaking (or tail wagging)	Tail shaking
Bathing	Bill dipping	Bill dipping
	Head dipping (also dip-diving or dipping)	Head dipping

Unritualized Comfort Movements

Wing-and-leg stretch
Two-wing stretch
Wing shuffling
Tail fanning
Scratching
Foot shaking

males. Instead, water-splashing noises and percussion sounds produced by drumming the lower mandible on the tracheal air sac, or other mechanically generated sounds made with bill, feet, or wings, seem to be used predominantly by males for generating auditory stimuli during social display.

A considerable number of comfort movements or comparable behaviors are performed by male stifftails during social display. These are sometimes performed in a seemingly functional manner and only questionably qualify as displays. Yet at other times they are clearly different in form and context from their functional (so-called autochthonous) comfort-movement counterparts. They often involve movements of the conspicuously colored bill and variably patterned head. Evolved ("ritualized" or "allochthonous") display versions and their comfort-movement counterparts may be conveniently summarized as shown in Table 9.

It may be seen that most of the usual comfort movements of anatids occur during social display of stifftails. Indeed these movements perhaps occur more frequently and with more diversity in this tribe during social interactions than in most other groups of ducks, although firm data supporting this subjective opinion are lacking. It is nevertheless possible, because of all the splashing and diving that

TABLE 10
Social Display Behavior and Functional Locomotory Movements

Movement	Ritualized Display	Functional Locomotory Behavior
Surface movements	Lurching *(Oxyura)*	Brief, rapid paddling
	Paddling kick *(Biziura)*	
	Travel-splash kick *(Biziura)*	
	Rushing, motor-boating *(Oxyura)*	Rapid paddling (surfing)
	Hunched rush *(Oxyura)*	Aggressive rapid paddling
	Swimming low and swift	Rapid paddling attack
	Flotilla swimming *(Oxyura)*	Coordinated swimming
Diving movements	Head-up, tail-cocked *(Oxyura)*	Diving intention (alarm)
	Wing-up, tail-up *(Heteronetta)*	
	Precopulatory head dip *(Oxyura)*	Actual dive: Head dipping
	Splash dive *(Biziura)*	Noisy escape dive
	Stealthy dive *(Oxyura)*	Silent attack dive
Takeoff movements	Head-up, tail-down *(Oxyura)*	Flight-intention posture
	Ringing rush *(Oxyura)*	Running takeoff

usually occur during stifftail display, that movements associated with water removal from the feathers are correspondingly increased.

Another group of displays is clearly derived from locomotory movements, consisting of the preliminary "intention movements" anticipatory of locomotion itself. Among these are such displays (or putative displays, in the case of intention movements) as outlined in Table 10. Several of these display postures have already been described in this chapter; most others are self-explanatory. Descriptions of all these movements, and illustrations of most, also appear in the various individual species accounts given in later chapters. Because of the varied amounts of ritualization in stifftail behavior, it is often difficult to distinguish ritualized displays from their nonritualized, functional counterparts, thus this listing may be incomplete.

MAJOR RITUALIZED DISPLAYS OF MALE STIFFTAILS

Besides those displays that are clearly derived from self-directed comfort movements or have obvious locomotory-based origins, male stifftails perform a variety of generally more complex and conspicuous displays that seem to represent important sex-specific and species-specific signals. It is these displays that tend to be the most eye-catching as well as the most audible, and the ones that often seem to be most closely associated with intense display activity. However, they need not always be interpreted as strictly "courtship" displays, inasmuch as many of them seem to be self-advertisement signals associated with territorial proclamation or may function in the related role of establishing dominance over other males and heterosexual control over or attraction of females. These displays will be briefly listed and described here for comparative purposes, although more detailed accounts can be found in the individual species descriptions in later chapters. The major male displays of the freckled duck and white-backed duck will also be briefly addressed here, simply to point out their strong differences from any known stifftail displays.

Males of the white-backed duck seemingly lack any of the elaborate social displays characteristic of stifftails (Johnsgard 1967, 1978). Sexual displays seem to be extremely inconspicuous, but possible pair-forming behavior in this species includes two birds swimming parallel while calling and chin lifting (Fig. 11A); mutual bill dipping may also be a part of this behavior. Pair bonding is strong and perhaps permanent, as in whistling ducks. An aggressive gaping, with a spreading of the flank and scapular feathers and an accompanying loud hiss (Fig. 11B, C) is a common male display when defending his nest or mate and is more that of whistling ducks than of stifftails. The white-backed ducks' postcopulatory display—a distinctive, mutual "step dance"—is exactly like that of whistling ducks.

Males of the freckled duck in breeding condition frequently perform one distinctly ritualized display, the "axel-grind" (Fig. 11D). This seems to serve both as a signal from a dominant male other males and as an advertising signal to females. The axel-grind display sequentially consists of a preliminary lateral head shake, which is followed by a forward head thrust with the bill tilted somewhat downward; a soft, squeaky, metallic vocalization accompanies this forward neck-stretching phase. The display ends with a final strong lateral tail wagging. None of the typically elaborate pair-forming male displays of *Anas* and closely related genera of dabbling ducks are evidently present. There are a few slight acoustic similarities between the axel-grind vocalization and that uttered by male black-headed ducks during their display, but these similarities are really quite superficial. The preflight behavior (Fig. 11E) of this strange species is somewhat *Anas* like, rather than being stifftail-like, and consists of a rapid, usually silent, head raising with the bill held horizontally, followed by a slow lowering of the head. Inciting behavior has not been observed, and copulation is performed without elaborate precopulatory or postcopulatory displays (Fullagar, Davey, and Rushton 1990).

In the black-headed duck, only one such display needs to be briefly discussed here. This is the "toad-call" (Weller 1968a) or "gulping display" (Johnsgard 1965a). The main advertising display of male black-headed ducks, it consists of throat inflation, rapid head pumping, and a repeated lifting of the folded wings, together with an accompanying weak vocalization (Fig. 12A). This strange display has no apparent direct counterparts among either the typical stifftails or with other waterfowl groups, including the freckled duck. Perhaps it originated as a way of forcing air out of the inflated cheek pouches, thereby producing a soft sound.

Several relatively complex male self-advertising displays occur in species of *Oxyura;* they mostly involve neck inflation and vocal or

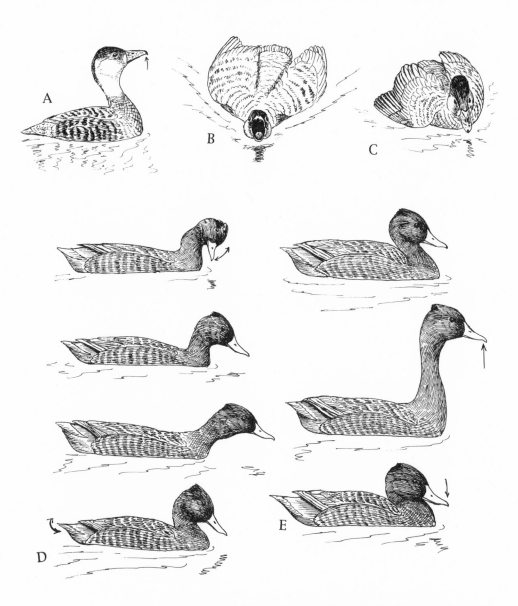

Figure 11. Social behavior of two purported stifftails. Male white-backed duck shown performing (A) chin lifting while calling, (B) intense threat display, and (C) low-intensity threat display. Also shown is male freckled duck performing (D) axel-grind display, and (E) preflight head movements. [After photos by Johnsgard (A–C) and sketches in Marchant and Higgins 1990 (D, E).]

mechanical sound production and are often seemingly associated with the most intense phases of social and sexual display.

The first of the complex male displays of *Oxyura* to be described in some detail was the "bubbling" display, which was first observed in the North American ruddy duck (Johnsgard 1965a) but later found

Figure 12. Some highly ritualized social-sexual displays of male stiff-tailed ducks: (A) toad call (gulping) by black-headed duck, (B) bubbling by North American ruddy duck, (C) sousing by Australian blue-billed duck, (D) territorial vibrating trumpet by maccoa duck, (E) kick-flap by white-headed duck, and (F) whistle kick by musk duck.

to occur in at least one of the two South American races as well. Basically it consists of inflating the neck region, cocking the tail, and then performing a series of several (usually three to eight) vigorous taps of the lower mandible on the upper breast, forcing air out from these feathers as a ring of bubbles. The display sequence is ended with a forward thrust of the head, maximum tail cocking, and a weak vocalization (Fig. 12B). The origin of this display is unknown, but in a manner somewhat like the toad call of the black-headed duck it seems to primarily function as a noise-producing device, in this case using the tracheal air sac as a percussion instrument.

The next of the major male displays of *Oxyura* to be described was the "sousing" display of the Australian blue-billed duck (Johnsgard 1966). It was later determined that the Argentine blue-billed duck possesses what is essentially the same display (Johnsgard and Nordeen 1981), as does the maccoa duck (Siegfried and van der Merwe 1975). In this display the bird typically begins in an erect-neck, cocked-tail posture, from which he swings his head downward and forward until it is just above the water; in this position he performs a variably extended series of compulsive jerky movements, as if he were choking (Fig. 12C). The male often terminates the display by shaking his head in the water, producing a lateral shower of water splashes.

The maccoa duck uses the sousing display rather infrequently; instead, his primary male advertisement displays seem to consist of the "vibrating-trumpet" call. There are actually two forms of this display, originally called the vibrating-trumpet call (here simply "vibrating trumpet") and the independent vibrating-trumpet call (here the "territorial vibrating trumpet"). In the vibrating trumpet the head and neck are first moved backwards and then extended forward as the neck region is greatly expanded; the tail is gradually cocked. A rather loud belching noise is produced until the bill touches the water (Fig. 12D). In the territorial vibrating trumpet the head is not initially moved backward, the bill does not actually extend so far as to touch the water, and the tail never reaches a vertical position when it is raised.

The male white-headed duck lacks all of these displays, including the bubbling display erroneously attributed to it in early accounts. Several complex displays are present, but the species' major male display is the "sideways hunch" (Matthews and Evans 1974). Unlike the other *Oxyura* forms, little if any neck inflation is evident in this species. Rather, the male holds his head low in a swimming position, with his tail flat in the water and vibrating. In this posture the male orients himself laterally toward a courted female and utters a mechanical "tickering purr" sound. A more elaborate sequence, the "kick-flap" (Matthews and Evans 1974), is much less frequently performed but is of special comparative interest because of its similarities to the various kicking displays of the male musk duck. In achieving this posture the male begins by lowering his bill into the water while his tail is cocked. Then he lowers his tail and quickly lifts his bill from the water, simultaneously shaking his head from side to side. He then strikes the water with his tail and simultaneously kicks upward with both feet, producing a splash of water (Fig. 12E). The display ends with the bird in a lateral "sideways-piping" posture, with the bill oriented toward the female as he calls repeatedly.

Finally, the male musk duck performs several displays that seem to form a hierarchical series of intergraded postures and intensity. Johnsgard (1966) has labeled them the paddle kick, plonk kick, and whistle kick, but Carbonell (1983) has called them collectively the "kicks series." The posture assumed in the whistle kick is characterized by an enlarged and turgid submandibular lobe, full inflation of the throat pouches, and extreme tail cocking (Fig. 12F). In this remarkable posture the feet are simultaneously and repetitively kicked upward at several-second intervals, sending showers of water to the side and rear. An extended series of loud whistling notes is produced at the rate of one per kicking movement. These same vocalizations are sometimes also uttered during performances of the plonk kick and paddle kick, but at substantially lower incidences (Carbonell 1983).

FUNCTIONS AND PHYLOGENY OF SOCIAL BEHAVIOR

Tinbergen (1954) pointed out that displays that have acquired a social stimulus or "releaser" function only secondarily evolved such functions, originating instead from nonsignal sources. He suggested that the social functions of courtship displays are fourfold: They synchronize the mating activities of the sexes; orient individuals toward one another; suppress nonsexual responses in females; and tend to be species-specific, to help preserve reproductive isolation.

The synchronization of stifftail sexual cycles is achieved by a combination of aggressive and sexual displays performed by the male, who in turn must respond in a manner that is at least partly sexual (Hinde 1955–56). If the female is ready to copulate, she may allow the male to approach, or alternatively she may peck at him. In *Heteronetta* the synchronization of reproductive activities may be aided by the constant observation of the host species' activities and associated nest-seeking behavior on the part of the female.

The second function of social releasers, that of interindividual orientation, is mainly achieved in stifftails through visual signals (bill and plumage coloring, posturing, etc.), but in some species acoustic stimuli associated with male displays, such as water splashing or loud vocalizations (e.g., musk duck, white-headed duck), may facilitate the orientation of females toward males or, alternatively, may help to space out competing males.

The suppression of nonsexual responses may be related to the positioning of males to females, since male stifftails either tend to begin display while maneuvering their body in front of and laterally to the female (e.g., black-headed duck, ruddy duck) or directly lateral to her (especially the white-headed duck). Additionally, white-headed ducks assume a hunched posture, which possibly represents a submissive, or at least nonaggressive, posture toward females.

The functional importance of reproductive isolation in the displays of stifftails is more questionable, since in most species there is little or no contact with other congeners, at least during the breeding season. Further, where the ruddy duck has come into contact with the white-headed duck, hybridization has occurred (in captivity and under natural conditions) in spite of the species' very different male signaling behaviors. It has been suggested (Tinbergen 1954; Morris 1957) that display diversity and thus reproductive isolation can potentially be gained by altering the speed of displays and the vigor of their individual components and by differential exaggeration or omission of individual displays or display components. It is very evident from their time-durational aspects that speed characteristics are probably a significant aspect of stifftail displays and that many of their displays are highly stereotyped as to their durational as well as postural aspects. However, there are some display sequences that have highly stereotyped components but quite variable overall durations, like the sousing display of the Argentine and Australian blue-billed ducks and the maccoa duck. These differences result from the number of pumping movements performed, which may in turn depend on the intensity of the stimulus or the male's internal state.

The vigor with which the display movements of stifftails are performed seems to be of great importance, and various writers have used the term *convulsive* to describe displays such as sousing and sideways piping. The kick-flap of the white-headed duck and the several kicking displays of the musk duck are even more extreme. Differential exaggeration occurs in the bubbling display of the ruddy duck as performed in the North American and South American races and in the degree and number of pumping and choking elements during sousing in the three species that perform it. Finally, rhythmic repetition is a major component of stifftail displays, as in the toad call of black-headed ducks, the bubbling of ruddy ducks, the sideways piping of white-headed ducks, and the various kicking displays of musk ducks.

The heterosexual functions of these highly rhythmic displays may include not only female-choice functions but may also result in a kind of mesmerizing effect on the female that tends to hold her in place long enough for the male to approach and perhaps attempt copulation. Similar seemingly female-mesmerizing displays occur in the male courtship displays of several other polygynous bird groups (Johnsgard 1994).

In some other duck groups, such as the eiders (Johnsgard 1964), the absence of some male displays among certain species in the group seemingly results either from the ritualization of such displays after the splitting of ancestral gene pools or from a secondary loss caused by selective pressures for species specificity. Both situations may also occur in the stifftails. For example, the ritualized wing flapping of the maccoa duck and the dab-preening of the Argentine and Australian blue-billed ducks may have evolved after the separations of these species, whereas the wing-lifting component (present in other

Oxyura as well as in *Heteronetta* and *Biziura*) is lacking in all three of these forms.

INTERSPECIFIC ETHOLOGICAL SIMILARITIES IN *OXYURA*

Table 11 a summary of the different display movements, and some of the male secondary sexual structures, that are part of stifftail social signaling, either as individual display units or display components. The masked duck is excluded from this summary because of the near-absence of detailed ethological information available for it. What seems to have made the male social displays of these stifftails most highly species-specific is the way in which different elements have been combined and variously ritualized to produce very different end results in a visual and acoustic sense, in a manner somewhat similar to that evident in the eiders (Johnsgard 1964). A summary of individual points needing explanation or warranting special comments follows.

Inflated neck and tracheal air sacs. These structures are associated with sound production and visual signaling (enlarged neck and throat) during male display. The esophagus is probably enlargeable through inflation in all species (not yet established for *leucocephala*); tracheal air sacs are present in at least some species, but their presence needs to be established for others.

Ruffled neck feathers. Erected neck feathers are present to varying degrees in all *Oxyura* species, except for *leucocephala*, during male display. It is sometimes difficult to separate neck inflation from feather erection, the two being normally associated with one another.

Feathered "horns." Paired, earlike crown feathers are present to a limited degree in males of *maccoa* but are best developed in *jamaicensis*, especially so in the North American race.

Wing lifting. A quick lifting of the folded wings appears during male display in all species except *vittata*, *maccoa*, and *australis* and is perhaps derived from the two-wing stretch, but is much more rapidly performed. The action in *jamaicensis* is associated with the neck-stretched-forward and extreme tail-cock posture at the end of the bubble display. In *leucocephala* it occurs during the kick-flap display. In the related genus *Biziura* it occurs during the kick-flap (whistle kick) and in *Heteronetta* during early phases of the toad call (Fig. 12).

Wing shaking. Wing shaking (also called wing ruffling) is present in *Heteronetta* and all *Oxyura* and is clearly derived from the corresponding comfort movement. It is performed by several species in the absence of other males, but *vittata* perform it while flotilla swimming and as part of a more complex fixed sequence (dab-preen, wing shake). In *Heteronetta* it precedes copulation, and in *leucocephala* it is part of sideways piping.

Tail cocking. This posture occurs in all *Oxyura* species (and *Biziura*) and probably originated from an extreme alert posture and es-

TABLE 11
Distribution of Male Stifftail Structures and Social-Sexual Signals

Display Component	Heteronetta	Oxyura jamaicensis	Oxyura vittata	Oxyura australis	Oxyura maccoa	Oxyura leucocephala	Biziura lobata
Social-sexual displays							
Swimming low and swift				X?	X		
Wing flapping				X	X		
Dab-preening			X	X	X		
Choking			X	X	X		
Swimming shake			X	X	X		
Sideways hunch						X	
Lateral posture			X	X	X	X	
Tail shaking			X	X	X?	X	
Tail vibrating				X		X	
Tail flashing/leading		X					
Bill-dip, head-flick		X	X	X	X	X	
Head dipping		X	X	X	X	X?	
Flotilla swimming		X	X	X	X	X	
Hunched rush		X	X	X	X	?	
Rushing/ring rush		X	X	X	X	?	
Sousing sequence			X	X	X		
Bubbling sequence		X					
Head pumping		X	X	X	X		
Cheek rolling		X	X	X	X		
Turning	X	X?					
Hunched posture	X	X	X	X	X		
Head-high posture	X	X	X	X	X	X	
Head shaking	X	X?	X	X	X	X	
Wing shaking/lifting	X	X	X	X	X	X	X
Tail cocking	X	X	X	X	X	X	X
Tail up and down						X	X
Foot kicking				X		X	X
Head stretched forward		X	X	X	X	X	X
Precopulatory displays							
Whistle kicking							X
Sideways hunch						X	
Neck jerking				X			
Bill flicking		X	X				
Mutual head pumping	X						
Neck/throat expansion	X	X	X	X	X	X	X
Feather ruffling	X	X	X	X	X	X	
Inflated tracheal air sac		X	X				
Inflated throat pouch	X						
Inflated esophagus		X	X	X			
Inflated cheek pouch							X
External features							
Feathered "horns"		X			X		
Cheeks variably white		X				X	
Body chestnut brown	1	X	X	X	X	X	
Bill bright blue	X	X	X	X	X	X	
Head mostly black	X	X	X	X	X	X	2

1. Brown, patterned with black, rather than entirely chestnut.
2. Upper head heavily spotted with blackish, not all black.

cape tendency. In *jamaicensis* it occurs during the bubbling sequence, in *vittata* and *australis* during sousing, in *maccoa* during sousing and the vibrating trumpet, in *leucocephala* during the head-high, tail-cock display, and in *Biziura* during the whistle kick.

Tail vibration. This action occurs in *leucocephala* as part of the male's sideways-piping display; in *australis* (where it is called "whisking"), it is often followed by a backwards kick and wing flapping.

Tail up and down. This action of raising and then immediately lowering the tail is clearly derived from the tail cock and is a component of the kick-flap of *leucocephala* and the various kick displays of *Biziura*.

Head high. This erect-neck display occurs in all *Oxyura* and also in *Heteronetta*, but not in *Biziura*. It no doubt originated from an alert posture and probably represents an escape-intention movement. In *Heteronetta* it occurs during the toad call; it is also present in the bubble display of *jamaicensis*; the sousing of *vittata, australis,* and *maccoa;* and the head-high, tail-cock of *leucocephala*. It is also an element in the head jerk of *vittata* and the vibrating trumpet of *maccoa*.

Head pumping. A vertical head-pumping movement during display has not been observed in *Biziura* or in *leucocephala,* but mutual head pumping is a part of copulatory behavior of *Heteronetta* (as in *Anas*). A more rapid (and probably nonhomologous), mostly vertical head movement is a major component of the bubbling display sequence of *jamaicensis* and the sousing sequences of *vittata, maccoa,* and *australis*. In *jamaicensis* there are from 2 to 8 head-pumping movements during the male's bubbling sequence, the mean number varying somewhat (from 3.0 to 6.7) with the several subspecies. In the latter three species the average number of head pumps during the sousing sequence is 2.6 (range 0 to 4) in *maccoa,* 4.2 in *australis* (range 1 to 11), and 6.8 in *vittata* (range 4 to 10). Its presence as a mutual precopulatory display in *Heteronetta* might suggest affinity with the dabbling-duck group, but no such comparable function is known to be present in *Oxyura* head pumping. It is possible that special mutual precopulatory displays were lost as pair bonding was replaced by promiscuous mating.

Head stretched forward. This is a major component of male *Oxyura* and *Biziura* displays. It occurs as the final phase of the bubbling display sequence of *jamaicensis,* is part of the head-jerking display of *vittata,* occurs during the vibrating trumpet of *maccoa,* and is present in the entire kick series of *Biziura*. In almost all cases it is associated with a vocalization, suggesting that a stretching of the trachea may be a basic part of its functional origin, as is also evidently the case in many neck-stretching displays of pochards, sea ducks, and dabbling ducks.

Choking. This convulsive display, also performed with the head and neck stretched forward, is present only in *vittata, australis,* and *maccoa,* where it occurs during sousing. It perhaps represents a ritualized form of the swimming shake, but a rhythmical element has been added, which is most elaborate in *vittata* and least in *maccoa*. Thus the mean number of chokes during sousing by *maccoa* is 2.2 (range 1 to 3), versus 5.1 (range 3 to 7) in *australis* and 9.0 (range 3 to 17) in *vittata*.

Bubbling sequence. This display complex, starting with vertical neck stretching and tail cocking, followed by a rapid tapping of the lower mandible on the breast and a final vocalization with the neck stretched forward, is unique to *jamaicensis*. The bill-tapping phase is certainly similar in some respects to choking, and perhaps both display elements evolved from a common source. Neither choking nor bubbling is present in *leucocephala*, but, interestingly, some first-generation male hybrids between *jamaicensis* and *leucocephala* produced at the Wildfowl Trust performed a fairly typical bubbling display (with about five head pumps followed by a single-note burp) whenever they were disturbed slightly (Hewston, pers. comm.). A second-generation male that is a back-cross to the white-headed duck has been observed performing the head-high, tail-cock display and a more rudimentary version of bubbling, consisting of only two head nods and a lurch forward (Hughes, pers. com.).

Neck jerking. This display of *O. vittata* is similar to head pumping, but it is much more rapidly performed, it has a greater range of jerking movements per sequence, and the "emphasis" of rapid movement is on the upward component rather than the downward. In that species it also is a major precopulatory display.

Rushing. This display, used by probably all *Oxyura* species, includes the hunched rush, which is a clearly aggressive approach toward other males with strongly raised scapulars. Since it is a functional threat/chasing/attack behavior, it need not be considered as representing a true sexual display. More sexually oriented rushes include a flying-running ringing rush (also called "hurried flight," "running flight," "display flight," "display run," and "flutter display"). The ringing rush (named for the ringing sound associated with it) apparently is used by males of several *Oxyura* species to approach females quickly or perhaps to draw acoustic and visual attention to themselves. A similar, but entirely surface-running, "rush" or "motor-boating" display seems to serve the same purpose in some species but does not involve any flapping of the wings.

Foot kicking. Simultaneous kicking of the feet, usually without accompanying submergence, may well be derived from a diving-escape intention movement. Unlike the feet of goldeneyes (*Bucephala*) and mergansers (*Mergus*), the feet of stifftails are not brightly colored, and the visual display effect is one primarily of water splashing rather than foot exposure. However, the sound produced by such kicking displays is considerably greater than occurs during normal diving or during the kick displays of the sea ducks just mentioned. Foot kicking may merge with loud and noisy "plop dives" in *australis* and, even more conspicuously, in *Biziura*.

Swimming low and swift (also called "swimming low and flat" and "stretch swimming"). This threatening posture, with the body and head low in the water in a submarinelike position, is evident only in *maccoa* and is certainly derived from an attack posture.

Flotilla swimming. This coordinated swimming activity is present in *Oxyura* but seemingly lacking in *Heteronetta* and *Biziura*. However, its presence may be strongly influenced by environmental conditions, such as the number of birds present at the local site, and its significance as a social signal (possibly even being primarily a coordinated alarm response) is uncertain.

Apparent male appeasement postures involving body or head orientation include (1) the sideways-hunch posture of *leucocephala*, which is a more ritualized version of the general hunching and partial withdrawal of the bill toward the body evident in several *Oxyura*, (2) lateral display (a lateral orientation of the body toward the female) and (3) turning display (alternate exposures of the male's flanks and tail to the female). Male stifftails sometimes approach females in an apparent appeasement posture (sideways hunch in *leucocephala*, general hunched posture with bill withdrawal and downward tilting in the others except for *Biziura*). When close to the female, males of *vittata*, *maccoa*, and *leucocephala* typically orient themselves to the side of the female (lateral display), from which their posturing and visual signaling can be easily seen, and which also may not be so aggressive an approach as would be a head-on orientation. However, the bill is always directly oriented toward the female during sideways piping by *leucocephala*, and the bubbling display of *jamaicensis* is oriented toward the female in the majority of cases. In *Heteronetta* and *jamaicensis* the males often constantly maneuver or jockey for a favored position directly in front of the female (a tactic sometimes called "lead swimming" or "tail flashing" in *jamaicensis*) or may also orient toward her side (typical turning display of *Heteronetta*).

Ritualized or displacement comfort movements are highly evident during *Oxyura* social display but are not obviously present during male display in *Biziura* and *Heteronetta*. They include (1) bill dipping, (2) head dipping (also termed "dipping" or "dip diving"), (3) head flicking, (4) head shaking, (5) dab-preening, (6) swimming shake (also termed "general shake"), (7) cheek rolling (also termed "rolling cheeks on back"), (8) tail shaking (also termed "tail wagging"), and (9) wing flapping. These activities are variably common during social display in all *Oxyura* species, and in at least *jamaicensis* and *maccoa* a combined bill-dipping and lateral head-flicking movement is a major component of male precopulatory display. Otherwise, the movements often differ only little if at all from their unritualized comfort-movement counterparts, suggesting that they have been only rather recently incorporated into stifftail social display.

The summary of display components in Table 11 places display or structural components that are more broadly distributed taxonomically toward the bottom, and those found in only a few species toward the top, with a few exceptions. It tends to support earlier classifications of the stifftails (e.g., Johnsgard 1978, 1979), except that it is now clear that leucocephala is rather sharply separated ethologically not only from jamaicensis with which it has often been affiliated in the past, but also from the other Oxyura species. Because of these distinctions, leucocephala's position in the genus is shifted in this book to a peripheral one, being placed last in the linear sequence of Oxyura species.

INTERGENERIC AND INTRAGENERIC RELATIONSHIPS IN *OXYURA*

The eggs of most species of Oxyura (all but *dominica*) are remarkably similar in being rather rough-surfaced, thus differing in texture from the smooth-surfaced eggs of Heteronetta and Biziura. However, the eggs do not shed any light on intrageneric affinities. Likewise, the ducklings and juvenal plumages of Oxyura, which closely resemble those of adult females, are not very helpful in this regard. However, the ducklings of leucocephala are notably dark in color and thus, at least superficially, tend to resemble those of Biziura. The ducklings of Heteronetta have more similarities to those of Anas than to Oxyura.

Adult female plumages of Oxyura fall into three general categories; those with heavy head striping (*dominica, leucocephala* and *vittata*), those with intermediate striping (*jamaicensis* and *maccoa*), and *australis*, which has very muted striping. These groupings do not clearly agree with any other line of evidence regarding classification.

Bill shapes in Oxyura are not helpful in establishing relationships either, although their measurements may be useful in characterizing individual species (Johnsgard 1967). The highly swollen bills of the white-headed duck and the maccoa duck are possibly associated with survival adaptations (salt removal) in brackish-water habitats. Swelling of the bill is considerably less evident in the other species.

With regard to their breeding adult male plumages, that of Heteronetta does not resemble one species of Oxyura any more than any other, and the sexually monomorphic plumages of Biziura are also not indicative of any specific affinities with any particular species of Oxyura. The white cheeks of adult white-headed ducks and of the North American (and Colombian) ruddy ducks appear to be a case of chance convergent evolution. This conclusion is based on the fact that the ontogenetic pattern of male plumage development in the white-headed duck is quite different from the patterns of geographic and individual variation found in ruddy ducks. (See Fig. 18.) Also, in contrast to all the other Oxyura species, the white-headed duck lacks a discernible nonbreeding and generally femalelike plumage. In this regard, as well as in its highly distinctive social display characteristics, the white-headed duck seems to have diverged considerably from the other Oxyura species.

The general plumage, morphological, and especially the behavioral similarities of the Argentine and Australian blue-billed ducks clearly suggest that they should be considered as closely related, indeed sibling species; the maccoa duck might also be added to this species pair as a slightly more distant form.

As mentioned in Chap. 1 the taxonomic status of *ferruginea* has been argued over for some time. Shortly after Delacour (1959) recognized that this form (together with the more northerly form *andina*) might best be regarded as subspecies of *jamaicensis*, Johnsgard (1965a) judged from Dr. Martin Moynihan's verbal descriptions of the male displays of *ferruginea* that it has a bubbling display much like that of *jamaicensis* and that this display is quite distinct from the displays of *vittata*. Livezey (1995) has recently recognized *ferruginea* as a full species, a sister group of *vittata* and *australis*, on the basis of cladistic analysis.

Johnsgard has since (1966, 1967, 1968b) suggested that strong similarities exist in the male sexual displays of *vittata*, *australis*, and *maccoa*. All of these species perform the complex sousing sequence, although with some differences in sequence length and number of repetitive display elements. In all of them the female uniquely tends to peck at the male while he is displaying. Clearly *maccoa* is behaviorally the most distinctive of these three, inasmuch as it exhibits both the swimming-low-and-swift and the vibrating-trumpet displays. Both of these displays, however, are related to the strong territoriality of male maccoa ducks, and as such the occurrence of these displays may be less important taxonomically than is the much less frequent occurrence of sousing. Johnsgard (1968) tentatively suggested that the maccoa duck may provide a kind of evolutionary transition between the display repertoires of *jamaicensis* on the one hand and *vittata* and *australis* on the other. This idea has received no additional support and now seems rather unlikely.

In any case, it is now fairly clear that *jamaicensis* is behaviorally closer to the *vittata*, *australis*, *maccoa* group than any of these forms are to *leucocephala*. However, this statement does not mean that *jamaicensis* should be regarded as having evolved recently from the South American populations of these species, nor does it support the possibility that *ferruginea* might have evolved directly from ancestral *vittata*, as suggested by Siegfried (1976c). It is certainly probable that the direction of evolution and speciation in *jamaicensis-andina-ferruginea* has been from the south northward rather than in the reverse direction, as was earlier suggested by Johnsgard (1965a). The white cheeks of the Northern Hemisphere *Oxyura* populations are thus evidently a derived rather than a primitive trait. Supporting this view is the fact that first-year *leucocephala* males have blackish cheeks, and a melanistic variant type of adult male *jamicensis* with black cheeks has been found (Todd 1979). The similarities of the white cheeks of *leucocephala* and northern populations of *jamaicensis* are unrelated and, therefore, lacking in taxonomic significance.

It might be hypothesized generally that the ancestral stifftails originated in the tropics of South America, or perhaps tropical Africa (assuming the distances between Africa and South America were still not very great during the early Cenozoic period), from which *Heteronetta* became isolated in southern South America rather early in the group's evolution. Contrarily, Gray (1980) suggested that stifftails could have invaded Asia from Australia and thence moved westwardly. This idea is certainly attractive in terms of explaining the present Australian distribution of *Biziura* but makes an early South American separation of *Heteronetta* much harder to fathom.

One might further hypothesize that ancestral *Oxyura* stock in tropical South America, or perhaps in Africa, initially produced *dominica* as a small tropical-lagoon specialist offshoot species in eastern or northern South America, following the group's development of lengthened tails and improved diving abilities. A second invasion, perhaps also from Africa, probably produced a more temperate or alpine-adapted population in the Andes, the forerunner of *ferruginea*. Northward expansion of this population initially then probably produced *andina* and finally gave rise to nominate *jamaicensis*. Why this more northerly population should have evolved whiter cheeks, larger tracheal air sacs, and a more prolonged bubbling display is still conjectural. However, it might well relate to the shorter pair-forming period available to stifftails for breeding in temperate climates—and the correspondingly stronger effects of sexual selection in favoring the most brightly colored and vigorous males. A similar explanation for the conspicuously white head color of the male white-headed duck may also apply. Thus the North American ruddy duck has the shortest mean egg-laying period of stifftails breeding at the Wildfowl Trust, and the white-headed duck has probably the second shortest. (See Fig. 10.) Nevertheless, male white-headed ducks at the Wildfowl Trust spend the greatest amounts of time is sexual-social display of all stifftails; the North American ruddy duck is only average in this regard. (See Table 3.)

Some time after giving rise to *ferruginea*, differentiation of the major *Oxyura* line occurred, which then probably expanded west into southeastern South America, producing the lowland-adapted *vittata*, and perhaps eastwardly into subtropical Africa, giving rise to ancestral *maccoa* stock. This or other *Oxyura* ancestral populations also presumably moved northward from Africa into the Sarmatic region of Eurasia (eastern Europe, western Russia), eventually giving rise there to *leucocephala*.

Such a hypothetical scenario for stifftail evolution still needs to explain the current presence of *Biziura* and *Oxyura australis* in Australia. Carbonell (1983) has suggested that *Biziura* may have been evolved when a common ancestor of *leucocephala* and *Biziura* split in vicinity of the Middle East, the former's parental stock then continuing north to colonize central Eurasia and the latter's progenitors moving southeast through southern Asia and eventually invading Australia from the north. This scenario would suggest that *Biziura* must be more closely related to *leucocephala* than the latter is to the other *Oxyura* species. If true, this would raise legitimate doubts about the generic validity of *Biziura* or, alternatively, might suggest that *leucocephala* should also be separated generically from the other *Oxyura* species.

The only other reasonable means by which *Biziura* might have invaded Australia is seemingly via Antarctica, when it provided a land connection between South America and Australia. This must have occurred much too long ago (late Mesozoic) to provide a likely dispersal scenario. However, a direct South America-Australia dispersal route would be a logical means of accounting for the strong behavioral similarities between the two present-day *Oxyura* species *vittata* and *australis,* which on the basis of ethological evidence seem to be more closely related to one another than are any other pair of *Oxyura* species. No other closely related ducks are shared zoogeographically between South America and Australia, but several (e.g., *Dendrocygna arcuata, Tadorna radjah, Anas superciliosa, and A. gibberifrons*) have definite zoogeographic connections with Indonesia. The present-day degree of separation between continental South America and Australia (about 8000 kilometers) strains credulity when one considers the probability of a direct and fairly recent Australian invasion of stifftail stock from South America. However, the Eurasian-breeding garganey *(Anas querquedula)* and northern shoveler *(A. clypeata)* have at various times both wandered as far south as Australia, and the North American blue-winged teal *(A. discors)* has accidentally reached there over even greater trans-Pacific distances of about 11,000 kilometers.

White-headed duck, pair swimming.

Reproductive and Population Biology

The reproductive biology of stifftails is an area of interest that has been only rather poorly studied for most species. This situation probably relates largely to the fact that breeding females tend to be elusive at that time of year, their nests are often well hidden, nest desertion is not uncommon following human disturbance, and most stifftail species breed in locations where biologists rarely do extensive fieldwork. Only for the North American ruddy duck may one say that the species' reproductive biology is exhaustively known, and although this species is probably fairly representative of the genus *Oxyura*, the corresponding strategies of *Heteronetta* and *Biziura* still need considerable work before it will be possible to thoroughly assess the reproductive biologies of these interesting species. The situation is especially true of *Heteronetta*, the only known obligatory social parasite among all the nearly 150 speciesof waterfowl and the only avian social parasite having precocial young. A single year-long field study (Weller, 1968a) provides the only currently available information for evaluating that species' reproductive adaptations. Even less is known of the overall breeding biology of the musk duck, although some useful information is available on its gonadal cycles as related to food availability, water fluctuations, and other environmental variables (Braithwaite and Frith 1969).

TIMING OF PAIR BONDING AND BREEDING

As noted in Chap. 2, sexual display among male stiff-tailed ducks at the Wildfowl Trust generally occurs over a very prolonged period, extending over virtually the entire time that males are in full breeding plumage in spring and summer. Whether or not this is reflective of the situation in nature for these species is questionable, since photoperiodic and weather differences are bound to be great for most species when comparing the situation in England with that in their native North American breeding areas. For example, at the Wildfowl Trust, at a substantially higher latitude but with a much milder winter climate than in interior North America, the male North American ruddy ducks are starting to develop breeding plumage in January, and all males are actively displaying not long afterwards, typically by the end of February or early March.

By contrast, in North American ruddy ducks most males are still in nonbreeding plumage during spring migration through the midwestern states (such as Nebraska) in March and April, and intense male display is not likely to be seen there until early May. In a northern California (Tule Lake) study area, Gray (1980) reported that among flocks of ruddy ducks on open water between late April and May 20, only 22 to 47 percent of the males had acquired their breeding plumages, although such plumages were present among 60 to 100 percent of the males already occupying nesting habitats during that same period. During late May and June the proportion of breeding-plumage males on open water increased to nearly 100 percent. By late August most of the males seen on open water were again in nonbreeding plumage, although all of those occupying nesting habitats remained in breeding plumage throughout that month.Females followed the males from the open-water habitats into the more vegetated nesting habitats, where display and pair formation then occurred. Sex ratios of the birds on arrival in open-water habitats averaged about 2.9 males per female (sample size, 1551 birds). The ratio of males to females became even higher during the prenesting (courting) period in the nesting habitats, then averaging 4.1:1 (sample size, 671 birds). During one of two years of study (1975), pair bonding was first observed late in the prenesting period (May 20 to 30), but in the following year it began about a month sooner, in late April.

Using marked birds, Gray (1980) found that most male ruddy ducks formed no pair bonds during the breeding season, but about one-quarter of them were associated with particular females for at least 2 days, and for as long as 24 days, and in one year three males were observed to form simultaneous multiple pair bonds. Females were more prone to bond with single mates, with more than twice the proportion of females forming a pair bond during both years than did males. However, females without pair bonds did reproduce. Two females were observed to form serial pair bonds during one year; in one case the female left with the victor after her male lost a fight, and in the other case a female who hatched a brood with one mate was later seen being attended by another male.

Most successful nests were initiated during late May in both years, although in 1976 the nesting season was shortened by increased water levels and flooding of nests in August and September. During the nesting period in June of 1975, about 70 to 74 percent of the females observed by Gray were judged paired, as compared with estimates of 78 to 87 percent the following year. Surprisingly, the highest observed levels of apparently paired females (89 percent) were observed early in the hatching period, rather than during the egg-laying period as might be expected.

These observations by Gray are notable for the high level of competition that must have occurred for mates among males, given the very unbalanced sex ratios. It is thus curious that no more than 85 percent of the females were ever observed to be apparently paired (the general averages for the late prenesting and nesting periods ranged from 62 to 83 percent), which suggests that perhaps a quarter of the females may remain unmated, even though they were all present in nesting habitats and presumably were breeding birds.

The other curious aspect of Gray's observations is that so many females seemingly remained "paired" after hatching. Probably this can be attributed to a prolonged sexual attraction of males to females (Johnsgard 1967; Kear 1970; Joyner 1975) or even to a social attraction of the males toward the ducklings, rather than to prolonged pair bonding as such.

TIMING AND PHYSIOLOGY OF NESTING

It is a general truism in ornithology that most species of birds have adaptively modified their breeding cycles so that their young will hatch at a time that is optimum in terms of available food supplies, inasmuch as the relative survival of the newly hatched young is one of the most important aspects of avian reproductive success. Thus the timing of pair-formation behavior—often apparently determined in waterfowl by photoperiodic changes that are consistent from year to year—may be considerably earlier than the timing of nesting and may also be under somewhat different timing mechanisms than those regulating the onset of nesting. However, at least at the Wildfowl Trust, there is additionally a clear relationship between the median laying day of each waterfowl species and the latitude of its native range. This relationship is consistent for all nontropical species of waterfowl (Murton and Kear 1976, 1978), which provides strong evidence that day length is a major timing factor for initiating breeding in nontropical species of waterfowl.

Some other obvious factors might influence the timing of breeding in tropical stifftail species, where annual photoperiodic changes are not very great, but where seasonally variable rainfall patterns may be well developed. Such variables might well include the relative local availability of suitable nesting cover (usually well-grown, emergent vegetation for stifftail nests) or an abundance of foods sufficient for females to incorporate prebreeding fat reserves to last through the egg-laying period.

Gray (1980) investigated the energetic aspects of reproduction in North American ruddy ducks and found that fluctuations in body weight associated with breeding were more marked in females than in males. Among males, average body weight increased 29 percent after the prebreeding molt in May and remained constant until the postbreeding molt in late August and September. However, females averaged a 66 percent increase in body weight prior to egg laying and lost over 25 percent of their body weight during the approximate week-long egg-laying period. By the end of the incubation period the female's body weight approximated the winter average. Most alterations in weight were associated with changes in stored lipids. Among males, lipids represented only about 5 percent of body weight during breeding, but among females almost 30 percent of their body weight was represented by lipid stores, and this prebreeding deposition of fats accounted for 47 percent of the total weight gained by females prior to laying. Energy budgets of females were highest during the prenesting period and lowest during incubation. Gray estimated that the costs of reproduction were 79 percent higher in females than in males and that the feeding requirements of females exceeded those of males by 67 percent, as measured by numbers of prey items consumed.

Carbonell (1983) also examined energy aspects of reproduction in stifftails. She concluded that breeding female stifftails must rely on daily food intake to obtain the energy necessary to lay a complete clutch. Stored fat reserves are used up during the incubation period, and periods off the nest for foraging are necessarily frequent. Females decrease their periods of time spent resting in the water as the amount of their time spent in the nest increases. The amount of foraging also increases during this period, with the foraging rate reaching its peak during the period of follicular growth, and especially on the day prior to the start of laying. Egg-producing females were found to spend a percentage of time foraging that was proportional to estimated energy intakes necessary for egg production. In various species this amount of foraging time was judged to range from as little as 9.6 hours per day in black-headed ducks to as much as 16.3 hours in maccoa ducks. All of the stifftail species studied suffered a considerable energy deficit during incubation, which only in the Argentine blue-billed duck was balanced by the surplus reserves acquired during the prebreeding and prelaying periods. The short but frequent periods off the nest for foraging may be facilitated by the ability of female stifftails to locate and ingest relatively large quantities of food in a short period.

Following breeding, the period of wing molt is an energetically demanding period, when the proteins associated with feather composition have to be replaced within a rather short period. Feather-protein requirements of ruddy ducks (feather-protein contents relative to total body mass) are variably greater than those protein requirements associated with reproduction, although reproduction costs are incurred over much shorter intervals (Hohman, Ankney, and Roster 1992). Protein reserves of ducks during the flightless period remain relatively constant, although large changes in two major protein structures breast and leg muscles, do occur. Thus, in North American ruddy ducks, overall body mass, protein content, and lipid content remain nearly constant during wing molt, but prior to the molt period there are substantial anticipatory losses (35 percent) of breast-muscle protein levels and considerable gains (15 percent) in leg-muscle proteins, apparently in correlation with the birds' greater needs for leg-powered locomotion during the flightless period (Hohman, Ankney, and Roster 1992).

NEST SITES AND NEST CONSTRUCTION

The nests of all typical stifftail species tend to be rather similar in appearance, materials used, and general measurements (Table 12). Only rarely is down present, and in such small quantities as to scarcely be considered a true lining. It would seem that the stifftails, together with some of the whistling ducks, are the only waterfowl that regularly breed without stripping down from their breasts and using it for a nest lining. Perhaps this curious fact is a reflection of the group's tropical origin, although the most northerly nesting species, the North American ruddy duck and the white-headed duck, also normally utilize very little nest down.

In spite of contrary reports from some authors (e.g., Palmer 1976; Cramp and Simmons 1977)—but in general agreement with others (e.g., Misterek 1974)—female stifftails, at least at the Wildfowl Trust, have never been observed to completely cover their eggs with down, grass, or other nesting materials. However, the eggs are also rarely exposed to overhead view, as the birds at the Trust typically nest in dense vegetational cover, which provides the female with stems that she bends over to build a partial dome above the nest, further obscuring the eggs from view.

Perhaps related to nest-building behavior is the fact that both sexes of some stiff-tailed ducks sometimes construct nestlike "platforms." These are not then modified by the females for nesting but may be used for loafing by prenesting adults (Siegfried, 1973a). Platforms may also be constructed later on by stifftail females for brooding their newly hatched ducklings (Marchant and Higgins 1990). Similar platform-building behavior has also been observed among male freckled ducks; in this species the platform started by the male may be adopted by the female for a nest site (Marchant and Higgins 1990).

TABLE 12
Nest Measurements* of Representative Stiff-tailed Ducks

	Sample Size, cm	External Diameter, cm	Internal Diameter, cm	Depth of cup, cm
North American ruddy duck	4	22.7	16.2	8.0
Maccoa duck	17	27.1	17.8	7.3
Argentine blue-billed duck	3	21.7	14.3	6.3
Australian blue-billed duck		ca. 40	ca. 20	ca. 7.5–10
White-headed duck	18	31.2	19.0	8.6
Musk duck	1	—	13.75	8.75

*All are means except figures for the Australian blue-billed duck and the musk duck.

Sources: Data on the Australian blue-billed duck and musk duck from Marchant and Higgins (1990). Other data from Carbonell (1983).

EGGS AND EGG-LAYING BEHAVIOR

Except for the black-headed duck, the masked duck, and the musk duck, all of which have relatively smooth-shelled eggs, the eggs of stifftails are typically rough-surfaced and are generally dull white to buffy white in color. They are also broadly oval to subelliptical (Table 13) and are relatively large as compared with the adult body mass of the female. This large egg size is correlated with the fact that not only are stifftail eggs thereby more resistant to chilling but also that the young are hatched in a highly precocial state and their good fat reserves effectively insulate them from the very beginning of post-hatch life for their highly aquatic existence (Lack 1967).

Eggs generally are laid by typical stifftails (*Heteronetta* are unreported) at an average interval of about 1.5 days per egg (Carbonell 1983; Marchant and Higgins 1990), although Helen Hays (cited in Bellrose 1980) reported an approximate daily egg-laying rate for four North American ruddy duck nests that she watched closely. The actual normal diurnal timing of egg laying is still evidently undetermined in this group. Most ducks eggs are laid at approximately daily intervals, and the eggs are usually deposited during early morning hours. The appreciably longer average egg-laying intervals of stifftails may well reflect their relatively large egg volume and associated laying stresses that do not permit daily egg production. In *Oxyura* species, the average fresh egg weight represents an estimated 16 to 23 percent of the adult female's weight (versus an estimated 20 percent in *Heteronetta* and 8 percent in *Biziura*); it is the largest relative average egg volume and mass of any waterfowl group (Lack 1967, 1968b; Siegfried, 1969b). Rohwer (1988) reported that among the stiff-tailed ducks a strong (0.87) correlation coefficient exists between egg mass and female mass for eight species of stiff-tailed ducks.

TABLE 13

Stifftail Egg Measurements, Weights, and Volumes*

	Length, mm	Breadth, mm	Mass, g	Volume, ml
Black-headed duck	55.3–61.7	40.2–45.1	57.9–65.1	50–65
	60.0 (21)	43.6 (21)	60.9 (4)	58.8 (8)
Masked duck†	53.7–55.6	40–41.6	52.0	48.5
	(5)	(5)	(est.)	(est.)
North American ruddy duck	57.9–70.9	41.9–49.6	60.5–85.7	60–75
	61.8 (189)	46.4 (189)	73.3 (56)	67.3 (28)
Maccoa duck	61.4–75.3	48.0–54.0	88.1–110.9	98–105
	68.6 (67)	51.2 (125)	99.7 (24)	96.25 (12)
Argentine blue-billed duck	61.1–74.3	45.7–49.7	74.8–89.6	75–80
	66.7 (67)	47.45 (67)	82.6 (21)	78.7 (15)
Australian blue-billed duck	64–72	46–51	84.0	77.8
	66 (62)	48 (62)	(est.)	(est.)
White-headed duck	60.4–75.9	46.3–53.3	77.4–105.8	80–100
	67.6 (185)	50.3 (185)	91.4 (37)	85.0 (24)
Musk duck‡	71–88	48–58	127	119
	79.0 (68)	54 (68)	(est.)	(est.)

*Weights are of fresh eggs from Wildfowl Trust, except for weights and volumes of masked duck, Australian blue-billed duck, and musk duck, which were estimated from egg measurements (c.f. Johnsgard 1972). Mean and sample sizes ( ) are shown below the ranges. Except as indicated, all data from Carbonell (1983).

†Earlier published masked duck egg measurements are questionable, but those from Cuban clutches seem reliable (Bond 1961; Wetmore 1965).

‡Musk duck egg measurements are those of wild birds (Marchant and Higgins 1990), since the few eggs laid the Wildfowl Trust seem atypically small.

The unusually large eggs of stiff-tailed ducks are not associated with corresponding increases in relative yolk quantities. The yolk contributes about 38 percent of the total egg mass, a proportion similar to that of other waterfowl groups. The rest of the egg mass in *Oxyura* consists of 49 to 51 percent albumen; the remaining 11 to 12 percent comes from shell weight (Siegfried 1969b). Presumably, the advantages of producing such unusually large eggs is that they allow stifftails to be hatched in a highly precocial state, which may be generally advantageous for allowing ducklings to begin prolonged dives for food at very early ages. (See Table 7 in Chap. 3.)

The large eggs of stifftails, and associated duckling precocity, may also have preadapted species such as the black-headed duck to become effective nest parasites (Weller 1968), or perhaps the larger-size eggs evolved after parasitic egg laying became adaptive among stifftails. The eggs of the black-headed duck are of interest in that they are not clearly modified for parasitic mimicry of host species such as coots. Although coot eggs are quite different in appearance from those of black-headed ducks, the latter's eggs are rarely expelled from coot

TABLE 14
Stifftail Clutch Sizes, Incubation Periods, and Duckling Weights*

	Clutch Size	Clutch Mass, g	Incubation, days	Duckling Weight, g
Black-headed duck	3–8 5(3)	305	22–25	30.5–41.5 36.9 (30)
Masked duck	6?	312	28?	28 (est.) 32.0†
North American ruddy duck	3–10 6.0 (37)	440	24–25	35.5–57 44.5 (35)
Maccoa duck	3–10 5.9 (39)	588	23–26	47.5–65 56.3 (31)
Argentine blue-billed duck	3–10 4.3 (18)	355	23–24	43–58 48.4 (24)
Australian blue-billed duck†	3–12 6.3 (10)	529	24–26	47–50 48.0 (4)
White-headed duck	3–10 5.8 (55)	530	24–26	47–65 56.0 (35)
Musk duck†	1–7 2.5 (50)	318	24	74 (est.)

*Means and sample sizes are shown below ranges. Clutch mass estimates are based on egg weights shown in Table 13. Unless otherwise indicated, data come from Carbonell (1983) from Wildfowl Trust captives.
†Data of Marchant and Higgins (1990), musk duck data from wild clutches; Australian blue-billed duck from captives. Estimates of masked and musk duck duckling weights are based on 58 percent of estimated fresh egg weight. (Mean ratio is determined from egg-to-duckling proportional weights of other stifftail species.) Actual masked duck duckling weight is average of two day-old ducklings from Trinidad (ffrench 1991).

nests (but may be buried in the nest), and various species of coots are seemingly the black-headed duck's most common hosts. The black-headed duck's eggs resemble duck eggs generally; they especially resemble those of rosy-billed pochards, a less common host.

Weller (1968a) noted that black-headed ducks "tend" to lay a single egg in each host's nest and suggested that nocturnal laying is a possibility, but not proven. Both of these traits would favor acceptance of the female's eggs. He also stated that the birds usually parasitized those nests in which the host had already laid three to five eggs. However, many eggs were also laid in nests where incubation had not only begun but sometimes was well underway.

CLUTCH SIZES AND THEIR ECOLOGICAL SIGNIFICANCE

Clutch sizes in the stifftails (Table 14), including parasitically laid clutches of *Heteronetta*, are generally relatively small compared with those of most ducks (Lack 1967, 1968b), but this characteristic is doubtless associated with their uniquely large egg size. Incubation periods

TABLE 15
Clutch-Size Variations of Stiff-Tailed Ducks, Especially in Relation to Female Ages

	Observed Clutch Size										
	1	2	3	4	5	6	7	8	9	10+	Mean
North American ruddy duck											
Age unknown (captive)	—	—	4	4	4	11	6	5	3	—	6.0
Age unknown (wild)	1	5	5	6	10	11	20	12	9	73	9.3
Maccoa duck											
1–3 years old	—	—	3	—	3	—	—	—	—	—	4.0
4+ years old	—	—	—	—	—	2	1	—	1	1	7.6
Total + age unknown	—	—	3	6	9	9	4	3	2	3	5.6
Argentine blue-billed duck											
1–3 years old	—	—	—	1	1	1	—	—	—	—	5.0
4+ years old	—	—	4	4	2	—	—	—	—	—	3.8
Total + age unknown	—	—	6	6	3	1	2	—	—	—	4.3
White-headed duck											
1–3 years old	—	—	1	—	1	2	3	—	—	—	5.8
4+ years old	—	1	—	—	6	5	—	1	2	1	6.1
Total + age unknown	—	1	5	6	13	14	9	4	3	1	5.7
All *Oxyura* species											
1–3 years old	—	—	4	1	5	3	3	—	—	—	5.0
4+ years old	—	1	4	4	8	7	1	1	3	2	5.3
Musk duck											
Age unknown (wild)	12	16	13	3	1	1	2	—	1	1	2.8

Source: Based on nests found at the Wildfowl Trust (Carbonell 1983), except for those of unknown-aged wild North American ruddy ducks (Joyner 1975) and musk ducks (Frith 1967).

consistently range from 22 to 26 days, a seemingly short period considering the large egg size.

In spite of earlier assertions to the contrary, there is no evidence that interspecific clutch-size variations in the stifftails are correlated with latitude (Rohwer 1992). Inter-and intraspecific clutch-size variations (Table 15) are surprisingly large among the stifftails studied in captivity by Carbonell (1983), and this same sort of variability is common among wild birds. At least part of the explanation would seem to be the frequency with which apparent single clutches are actually the result of two or more females laying in the same nest. Among 147 North American ruddy duck nests in Utah, Joyner (1975) found that the modal clutch size (with 20 nests) was 7. However, almost half of the total clutches (48 percent) had 10 or more eggs present (maximum of 23), suggesting that many of these nests must have been influenced by intraspecific parasitism.

Clutch sizes in *Oxyura* were found by Carbonell (1983) to be weakly positively correlated with the age of the laying female in two of three *Oxyura* species that were tested statistically. The information summarized in Table 15 (reorganized into larger age-class groups from Carbonell's original data) might suggest that older females may tend to produce slightly larger clutches than do younger ones, but it is also possible that they might only produce more variable ones. This could be the case if the older females are more prone to being nest parasites than are younger ones. Among the four species of *Oxyura* breeding at the Wildfowl Trust, individual egg-breadth variations were not found to be correlated with clutch-size differences, except for the North American ruddy duck where there was a significant inverse statistical correlation.

Relative egg mass is, of course, at least partly related to the ratio of total clutch mass to adult female mass. This inverse egg size to clutch size relationship was used by Lack (1967, 1968b) in his proposal that the clutch size is limited by the female's ability to produce eggs as affected by her energy reserves at the time of laying. This relationship was reexamined and recalculated by Rohwer (1988), who judged this relationship to be weaker than believed by Lack and estimated that clutch-size variations might account for only about 13 percent of interspecific egg-size variations among waterfowl. Using different statistical reasoning, Blackburn (1991) believed this influence to be more than twice as great, namely, 29 percent. In any case, using total estimated clutch-size mass as an estimate of female reproductive investment is perhaps a more useful statistic than egg mass if overall physiological stress of laying is to be considered. Among waterfowl generally, the clutch weight is inversely related to adult female weight, as is also the case for individual eggs. Among dabbling ducks and perching ducks the percentage of clutch mass relative to female mass varies from about 50 to 90 percent, for pochards from about 45 to 70 percent, and for sea ducks from about 25 to 110 percent (Johnsgard 1973). Among the stifftails the estimated proportion of average clutch mass relative to adult female mass is 21 percent in *Biziura*, 62 percent in *Heteronetta*, and from 62 to 90 percent in *Oxyura*. (See Table 16.) Briggs (1988) made comparable clutch mass estimates for four stifftails, finding similar proportions for the musk duck (21 percent) and North American ruddy duck (75 percent), but much lower estimates for the Australian blue-billed duck (42 percent) and maccoa duck (49 percent), mostly owing to smaller estimated average clutch sizes. Thus these percentages are at best rather crude figures, since clutch sizes in stifftails are especially difficult to judge accurately because of the birds' dump-nesting and parasitic-laying tendencies. Furthermore, all female ducks tend to put on substantial additional body weight (up to about 50 percent) just prior to egg laying. This fact is usually not taken into account (as it is not here) when calculating female-to-egg or female-to-clutch mass ratios, inasmuch as few female weights are available for this short but critical period.

Keeping such limitations in mind, one might tentatively conclude that the estimated clutch-to-hen mass ratio is seemingly low in the *Oxyura* species that are believed to be relatively monogamous (Argentine and Australian blue-billed ducks) but appears to average

Black-headed duck, pair in breeding condition. Photo by P. Johnsgard.

Black-headed duck, pair in breeding condition. Photo by P. Johnsgard.

Masked duck, male in breeding plumage. Photo by Mike Krzywonski.

Masked duck, adult female. Photo by Dr. Humberto Alvarez.

North American ruddy duck, group of adults in breeding plumage. Photo by P. Johnsgard.

North American ruddy duck, breeding male in partial tail-cocking display. Photo by P. Johnsgard.

Peruvian ruddy duck, breeding male. Photo by P. Johnsgard.

Maccoa duck, breeding male. Photo by P. Johnsgard.

Maccoa duck, male performing territorial vibrating-trumpet display. Photo by P. Johnsgard.

Argentine blue-billed duck, pair in breeding plumage. Photo by P. Johnsgard.

Argentine blue-billed duck, breeding male. Photo by P. Johnsgard.

Australian blue-billed duck, breeding pair. Photo by Frank Todd.

White-headed duck, pair in breeding plumage. Photo by P. Johnsgard.

White-headed duck, male performing head-high, tail-cock display. Photo by P. Johnsgard.

Musk duck, pair in breeding plumage. Photo by P. Johnsgard.

Musk duck, male performing whistle-kick display. Photo by P. Johnsgard.

higher in such apparently polygynous forms as the white-headed duck, maccoa duck, and North American ruddy duck. In these latter species larger clutches might be advantageous to counteract the possibly lowered hatching success associated with an absence of male protection from harassment during incubation. However, the ratio is also extremely low in the promiscuous musk duck, which has one of the smallest clutches of all waterfowl. This remarkably small clutch (of only two or three eggs) has been attributed to this species' unique (among anatine ducks) parental trait of the female directly feeding her dependent young and the dependence of young musk ducks on such parental feeding for much of their prefledging period (Kear 1970).

INCUBATION PERIODS AND THEIR ECOLOGICAL SIGNIFICANCE

Incubation behavior by stifftails has been described by Siegfried, Burger, and Caldwall (1976) for North American ruddy and maccoa ducks; their observations generally agree with Carbonell's at the Wildfowl Trust. All the *Oxyura* species there spent about 91 to 93 percent of the daytime hours on the nest and about 4.2–5.8 percent feeding. The nesting bouts lasted 4 to 5 hours, with intervening break periods of about 15 minutes. No nighttime observations were made, but Siegfried et al. (1976a) observed that in the two species they studied attentiveness was higher at night, though some periods were spent off the nest then as well, with about the same amount of time spent off the nest at night as during the day.

Incubation periods in the stifftails are surprisingly uniform (Table 14), considering the substantial differences in egg volumes occurring in the group. At least among the waterfowl species breeding in temperate and arctic regions of North America and Europe, there is a straight-line statistical relationship between egg weight and average incubation periods (Owen and Black 1990). All of the stifftails (except for the masked duck, whose estimated 28-day incubation period is questionable) fall below the mean correlation line (having shorter than expected incubation periods) that Owen and Black plotted for the Anatidae. The species that they plotted—no stifftails were included—having the greatest divergence toward short-than-expected incubation periods were two high-arctic geese (*Anser caerulescens* and *A. rossi*), which evidently evolved in and are ecologically adapted to nesting in regions with very short breeding seasons. Stifftail incubation periods are not of such extremely short durations as are those of either *A. caerulescens* or *A. rossi*, but they clearly fall below the average. Evolutionary reasons for relatively short incubation periods in stifftail ducks are difficult to hypothesize; the birds mostly nest in areas where the breeding seasons are generally long. Perhaps the special water-level-dependent conditions of ideal nest sites in stifftails make a brief incubation period especially desirable.

DUCKLING WEIGHTS AND THEIR ECOLOGICAL SIGNIFICANCE

The weight of the newly hatched duckling, relative to that of the adult female, is also of considerable ecological interest, as it might reflect such adaptations as the relative resistance of the duckling to chilling; its independence of its mother for protection; its ability to swim, forage, and dive soon after hatching; and its ability to escape from predators. Among stifftails, the musk duck has relatively the smallest-sized offspring at hatching (4.8 percent), which correlates with the very high level of prolonged parental care exhibited in that species. In the parasitic black-headed duck the estimated relative duckling-to-hen mass (6.5 percent) is within the mass range shown for all six *Oxyura* species (5.6 to 9.9 percent), suggesting that it is no better preadapted to a parasitic existence at hatching, in terms of its probable precocity, than are most of the typical stifftails. However, these figures should be considered in relationship to the relative hatchling-to-hen weights of other groups of waterfowl. Smart (1965) summarized a considerable amount of weight data for newly hatched waterfowl. Using these weights, plus those given by Johnsgard (1978) for adult females, it is possible to estimate comparable chick-to-adult ratios. Thus, for ten species of *Anas* the average hatchling-to-hen ratio is 3.5 percent (range 2.6 to 4.8 percent); for five species of pochards it is 3.2 percent (range 2.8 to 3.7 percent); and for six species of sea ducks it is 4.3 percent (range 3.2 to 5.7 percent). It is clear that, except for the musk duck, relative duckling weights at hatching are consistently higher in the stifftails than they are in any of the other major duck groups. In that sense, perhaps all of the *Oxyura* species are preadapted to a semiparasitic mode of reproduction because of their general capabilities for survival at hatching. However, their fledging periods are relatively long (Table 16), which means that the young birds must generally rely on diving for escape during their seven- to ten-week fledging periods. Additionally, such long fledging periods may effectively limit the extension of stifftail breeding ranges into areas having short frost-free seasons or into warmer habitats that are prone to dry up late in the summer season.

DUMP-NESTING AND PARASITIC EGG LAYING

In a review of known cases of conspecific brood parasitism, Rohwer and Freeman (1989) implicated all seven species of *Oxyura* as well as the musk duck in this behavior. Salyer (1992) noted that both conspecific and interspecific brood parasitism are known to occur in virtually all stiff-tailed ducks. However, separation of nonadaptive dump-nesting (in which the clutches are typically not incubated by any of the participating females) from adaptive nest parasitism (in which the clutches typically are incubated by the "host" female) is difficult if not impossible when considering the stiff-tailed ducks. Car-

TABLE 16

Stiff-Tailed Duck Clutch and Duckling Weights Relative
to Adult Female Weights and to Fledging Periods*

	Female Mass, g	Clutch/ Hen Mass, %	Duckling/ Hen Mass, %	Fledging Period, days
Black-headed duck	565	62	6.5	ca. 70
Masked duck	346	90	9.2 (est.)	45+
North American ruddy duck	513	86	8.6	52–66
Maccoa duck	677	87	8.3	63–70
Argentine blue-billed duck	560	64	8.6	56–63
Australian blue-billed duck	852	62	5.6	?
White-headed duck	593	84	9.4	58–70
Musk duck	1551	21	4.8 (est.)	?

*Mean egg weights from Table 13. Estimated clutch masses based on Table 14. Adult (nonbreeding) female weights from Johnsgard (1978) and this book. Fledging periods mostly from hand-reared captives (Carbonell 1983).

bonell (1983) observed that almost all the cases of known intra- or interspecific nest parasitism among four *Oxyura* species at the Wildfowl Trust occurred during the beginning of the breeding season. Joyner (1975) similarly found that the incidence of parasitized host nests rose at a slightly more rapid rate than that of wild North American ruddy duck nests, also indicating that most of the parasitic eggs are laid early in the breeding season. Joyner judged that it was nesting ruddy ducks, rather than transient or nonnesting birds, that were responsible for most of that species' nest parasitism in his study area. It may well be that at least in the North American ruddy duck and other parasitically inclined *Oxyura* species many of these eggs are deposited as a result of poor synchronization between behavioral nest-construction tendencies and physiological egg-laying cycles, rather than because of any biological advantages in parasitism as such. Given this consideration, it becomes difficult if not impossible to separate adaptive (or "purposive") parasitic egg laying from nonadaptive (or "accidental") dump-nesting of eggs in other birds' nests, simply because a female's own nest has not yet been completed.

This uncertainty causes a reconsideration of the significance of obligatory parasitic egg laying in the black-headed duck (Tables 17 and 18). Among captive birds at the Wildfowl Trust (Table 17), the common gallinule *(Gallinula chloropus)* was the most frequently parasitized species during Carbonell's study, although rosy-billed pochards were a close second, and the number of black-headed duck eggs present per nest averaged somewhat higher in the latter species' nests. The Argentine blue-billed duck (unreported as a host species in the wild) received only small attention, as did the mallard, Cape shoveler *(Anas smithi)* and the North American black duck *(Anas ru-*

TABLE 17

Incidence of Egg Parasitism of Black-Headed Ducks
in Captivity and under Natural Conditions

	No. of Nests Parasitized	Total Parasitic Eggs	Range of Eggs/Nest	Mean no. of Eggs
Captives (Wildfowl Trust)				
Common moorhen	12	37	1–8	3.1
Rosy-billed pochard	9	33	1–8	3.7
Mallard	5	13	1–7	2.6
Argentine blue-billed duck	5	5	1	1.0
Abandoned nests	2	5	1–4	2.5
Cape shoveler	1	3	3	3.0
North American black duck	1	1	1	1.0
Wild birds (Argentina)				
Red-gartered coot	31	56	1–8	1.8
Red-fronted coot	11	14	1–3	1.3
White-winged coot	8	10	1–2	1.2
White-faced ibis	32	36	1–2	1.1

Sources: Data for captives from Carbonell (1983); data for wild birds from Weller (1968a).

bripes). Since all black-headed duck eggs were collected whenever they were discovered in order that they might be artificially incubated, no information on relative host acceptance or hatching success with these atypical host species could be obtained.

The only information on egg-laying behavior and associated hatching success of black-headed duck eggs under natural conditions comes from the work of Weller (1968a). He provided a list of 12 species that have reportedly served as hosts; the species seemingly include virtually all medium to large (gull to swan) marsh-nesting birds that typically nest in fairly dense emergent vegetation, regardless of the size or color of their eggs. Parasitism incidence was highest for those species that nest in dense marshy areas, including especially the red-fronted coot *(Fulica rufifrons)* (55 percent of 133 nests) and the rosy-billed pochard (83 percent of six nests). Red-gartered coots *(F. armillata)*, nesting in more open vegetation, were parasitized less frequently (15 percent of 51 nests), and the white-winged coot *(F. leucoptera)* was evidently also only infrequently exploited. Weller believed that black-headed ducks selectively lay in active rather than deserted nests. However, coots often bury the distinctively unpatterned duck eggs, suggesting that they may recognize them as being foreign introductions. Weller also judged that 53 percent of 76 black-headed duck eggs he found in one study area had been deposited in nests during the host's (red-fronted coot) egg-laying period. Additionally, 35 percent were probably laid about 2 to 5 days late in optimum deposition time but still might have hatched (the coot eggs

TABLE 18

Relative Reproductive Success of Parasitic Stiff-Tailed Duck Eggs
as Compared with Those of Their Host Species

	Total Nests	Total Eggs	Nesting Success, %*	Hatching Success, %*
Black-headed duck†				
Red-fronted coot	58	—	83	—
Black-headed duck	40	73	26	22
Red-gartered coot	49	—	57	—
Black-headed duck	8	8	(0)	(0)
Rosy-billed pochard	6	—	(13)	—
Black-headed duck	5	5	(17)	(9)
North American ruddy duck				
Utah‡				
Cinnamon teal	11	71	—	62
Ruddy duck		28	—	11
Redhead	10	62	—	3
Ruddy duck		17	—	18
Northern pintail	3	14	—	29
Ruddy duck		12	—	92
Mallard	2	11	—	18
Ruddy duck		11	—	36
Wisconsin§				
American coot	5	0	(0)	(0)
Ruddy duck		5	(0)	(0)
Mallard	3	19	33	36
Ruddy duck		4	(33)	(25)
Redhead	1	8	(100)	(75)
Ruddy duck		3	(100)	(66)
Ring-necked duck	1	5	(0)	(0)
Ruddy duck		2	(100)	(100)
Pied-billed grebe	1	0	(0)	(0)
Ruddy duck		2	(0)	(0)

*Percentages shown in parentheses are those based on sample sizes of less than 10.
†Data of Weller (1968a). Nest totals for host species also include unparasitized nests. (Nesting success for 15 unparasitized red-fronted coot nests was 87 percent.)
‡Data of Joyner (1975), excluding nests also parasitized by redheads. Nest totals for host species are for parasitized nests only. Hatching success rates for unparasitized host species' nests were cinnamon teal 58 percent (1271 eggs), redhead 37 percent (508 eggs), northern pintail 66 percent (125 eggs), and mallard 65 percent (66 eggs).
§Data of Misterek (1974). Nest totals for host species are for parasitized nests only. The nests of the coots and grebe were old nests, in which no host eggs were deposited.

requiring 24 to 29 days of incubation). Finally, 14 percent were laid so late that there was no chance of hatching and duckling survival.

Considering that, like other coot species, the red-fronted coot probably incubates fairly steadily from the time that the first egg is laid in the nest, it is surprising that the black-headed duck is able to introduce any of its eggs into the nest at all. As mentioned earlier, Weller believed that nocturnal egg laying is a possibility (assuming that less nest constancy by coots might occur at night); otherwise the duck would have to be able to lay its eggs during those periods of daylight when the coot leaves its nest briefly to forage. It seems unlikely a female black-headed duck would be able to displace a coot forcibly from its nest long enough to lay an egg without interruption. However, like other ducks, rosy-billed pochard females do not tend their nests until incubation begins with clutch completion, so introduction of parasitic eggs into pochard nests would probably be much easier. Additionally, the eggs of these two duck species are so similar in size and color that is highly unlikely that any black-headed duck eggs could be recognized as foreign by the host rosy-bills.

HATCHING AND NESTING SUCCESS

The North American ruddy duck is practically the only species for which any adequate information is available on hatching success (percentage of eggs in initiated nests that hatch) and nesting success (percent of nests initiated that result in at least one egg hatching). Many nesting studies have been conducted on ruddy ducks under natural conditions across this species' breeding range in North America. Some of the results of these studies, in terms of relative reproductive success, are summarized in Table 19. Mean clutch-size estimates for wild ruddy ducks average about 8 eggs, ranging between 7.6 and 9.5 eggs. Such figures must be interpreted in the light of this species' strong tendency for engaging in parasitic nesting, as well as the "dump-nesting" of eggs in common nests by two or more females. Thus these estimates are generally not out of line with Carbonell's estimates of a normal 6-egg clutch for ruddy ducks, assuming that an average of 1 to 3 eggs may be added to most nests as a result of such collective female efforts.

Average reported nesting success in North America ruddy ducks (Table 19) is relatively high as compared with wild ducks generally; their aquatic nesting sites probably tend to reduce the incidence of losses to terrestrial predators. Aerial predators such as corvids evidently take relatively few eggs (Bellrose 1980). Of course, the selection of water-dependent nest sites also increases the probabilities of losses resulting from flooding or, alternatively, from rapid drops in water levels that may leave nests stranded on dry land and cause nest desertions. Bellrose (1980) estimated that nest desertion was the most common cause of ruddy duck nest losses (60 percent), followed by flooding (25 percent), unknown factors (10 percent), and avian egg predators such as crows and magpies (5 percent).

By comparison, a pooled sample of 1715 canvasback (*Aythya valisineria*) nests from various nesting areas had an average nesting success of only 46.2 percent, with raccoon predation the most serious single factor in nest and egg loss, followed by skunk and avian (usually corvid) predation; nest desertion was a variable influence. Studies involving a grand total of 1054 redhead (*A. americana*) nests show

TABLE 19

Some Reported Nesting and Hatching Success Rates of Wild North American Ruddy Ducks

Reference	Total Nests	Total Eggs	Mean Clutch	Nesting Success, %	Hatching Success, %*	Mean Early Brood Size†
Williams and Marshall 1938	19	158	8.3	—	52	4.3
Low 1941	71	546	7.6	73	69	7.3 (52)
Steel, Dalke, and Bizeau 1956	18	77	7.7	56	49	6.5 (300)
Reinecker and Anderson 1960	36	332	9.4	55	73	5.1 (107)
Misterek 1974	30	286	9.5	53	44	7.9 (17)
Joyner 1975	152	1419	9.3	34	31	8.5 (83)
Siegfried 1976a	40	—	—	70	—	—
Bellrose 1980‡	356	—	—	69.6	—	—
	312	—	8.05	—	—	—
	130	—	—	—	—	5.7 (962)

*Hatching success shown for successful nests only.
†Number of broods in sample shown in parentheses.
‡Data of Bellrose are based on summaries of several field studies, including some of those cited in this table.

a similar overall nesting success of 52 percent, with nest desertion (usually caused by intruding parasitic redhead females) a major loss factor, predatory mammals (raccoons and skunks) and birds a secondary destructive factor, and flooding sometimes also important (Bellrose 1980). Canvasbacks and redheads nest in much the same habitats and have nest-site characteristics similar to those of North American ruddy ducks, although they are both more prone to nest on dry land at times, especially the redheads.

The mean brood size of newly hatched North American ruddy duck young approaches 6.0 in most field studies (Table 19). This estimate is about 25 percent below that of the species' mean natural-condition clutch size of about 8 eggs, suggesting that there are probably only moderate losses of eggs among successfully incubated ruddy duck nests. Contrarily, estimated hatching success rates are usually substantially lower than are most estimates of nesting success. This probably means that, when some eggs are lost, it is likely that the entire clutch will be lost or abandoned.

Ruddy duck brood counts during the first week or so following hatching are quite high (5.69 for 962 still-downy broods, as reported by Bellrose 1980) and indicate that very little mortality probably occurs during that critical period, when most other duck species are typically suffering their highest duckling mortality. Presumably this can be attributed to the relative precocity and vigor of newly hatched *Oxyura* young. Yet counts of older broods show significantly smaller brood sizes. A mean of 4.43 young for 164 fully feathered broods was reported by Bellrose (1980), representing a posthatching reduction in average brood size of about 25 percent. However, this apparent brood reduction is possibly not so much a reflection of additional duckling mortality as it is the strong tendency for ruddy duck broods to break up as rapidly as the young birds become independent of their mothers.

FEMALE BROODING AND BROOD-RELATED BEHAVIOR

Good observations of brood-related behavior in stifftails have been provided by Joyner (1975), Siegfried (1977), Gray (1980), and Hughes (1992). Joyner (1975) observed that female North American ruddy ducks accomplished both interspecific and intraspecific brood defense by means of agonistic displays and actual aggression. Aggressive behavior by the female includes swimming near the intruder with the body plumage fluffed, lifting the folded wings and scapulars, stretching the head forward while hissing and gaping, and loud splashing with the feet (Marchant and Higgins 1990). Injury feigning by females is quite rare but, contrary to the experiences of some observers (e.g., Joyner 1975) it is not entirely lacking in stifftails (Siegfried 1977; Gray 1980). Communication between female and brood is achieved mostly through visual signals such as the female's alert postures (head raising, tail cocking, and general plumage flattening). Occasionally, auditory signals are also used by brooding females, especially when the young are visually separated from their mother, but stifftails of all ages are seemingly far less vocal than most ducks and fewer vocal signals are associated with their pair and family bonding (Siegfried 1976c). North American ruddy ducklings produce both distress and "contentment" calls, to which females typically respond. Disturbed broods tend to cluster near the female, but the ducklings respond to swoops by aerial predators by diving, simultaneously producing a loud splash that is unique to this situation. Both females and ducklings respond to avian predators by performing a weak version of the male's bubbling display (suggestive of its basically hostile origins). Calls are used to regroup scattered ducklings, whereas visual displays are used to stimulate specific brood responses, such as the possible use of tail cocking as a signal to follow in leading the brood away from possible danger.

Brood behavior in the presence of potential predators varies according to the age of the ducklings and the location of the encounters (Joyner 1975). Brooding female ruddy ducks tend to be highly aggressive and direct most of their intraspecific threats toward attending males, with progressively smaller percentages directed toward other males, ducklings belonging to other broods, and other females (Hughes 1992). Interspecific aggressive interactions are directed toward a wide variety of species. Even considerably larger species such as coots, mallards, and shelducks (*Tadorna tadorna*) are often dominated by these surprisingly domineering birds (Joyner 1977a; Hughes 1992).

During the first few days after hatching the female may return with her brood to the nest for brooding, or she may build a platform of vegetation near the nest, to be used for brooding and loafing. Unlike many species of North American ducks, broods of ruddy ducks rarely travel overland (Evans, Hawkins, and Marshall 1952).

Gray (1980) observed that males often accompanied females with young broods (present with 26 to 36 percent of early-hatching broods versus 9 to 11 percent of late-hatching broods). However, no male defense of broods was ever observed by Gray, nor has it ever been reliably confirmed by others (e.g., Johnsgard, Carbonell, Joyner) in spite of repeated statements to the contrary in earlier literature. Thus Joyner reported that during 163 interspecific aggressive female encounters, not once did the drake intervene on behalf of the brood. The majority (61 percent of 102 encounters) of interaspecific hostile encounters involved intrusion by courting males, and most others involved hostile female-to-female interactions. Joyner also observed that hens became increasingly intolerant of the accompanying drakes and gradually shifted from gaping to physical aggression, to which the males responded by retreating and eventually abandoning their associations with the hen and brood. Gray observed that few females remained with their broods after the latter were more than an estimated two to three weeks of age, or about five to six weeks prior to their time of actual fledging. Similar early abandonment of broods has been observed generally; Joyner estimated that brood abandonment occurred when the young were about four to five weeks old. Siegfried (1977) reported that females remained with their broods for at least four weeks, while Carbonell noted that a brood she was watching had become nearly independent with 24 days of hatching.

DOUBLE-BROODING AND RENESTING

Palmer (1976) reported that double-brooding (two successive clutches hatched and reared in a single breeding season) may occasionally occur in areas having longer breeding seasons than occur at the northern parts of the ruddy duck's breeding range, such as in Utah, which would perhaps help explain the typically early abandonment of their broods by stifftail females. However, no evidence of double-brooding or even renesting was reported by Siegfried (1976c) or Evans, Hawkins, and Marshall (1952) among ruddy ducks nesting in southern Manitoba, where the breeding season is appreciably shorter than is the case in California. Misterek (1974) found a bimodal pattern of laying by ruddy ducks in Wisconsin but believed that the first peak (in late May) probably represented laying activities of adult females, and the second peak (in late June) was caused by nesting of first-time breeders.

There is no good evidence from Joyner's (1975) data on nesting spans and hatching dates in Utah—or in Gray's (1980) comparable data from California—that suggests the existence of any second broods, which would necessarily have to hatch at least six to seven weeks later than the first broods, assuming the first were tended for at least three weeks after hatching and a second clutch was begun immediately thereafter. Joyner's data on nesting phenology for 113 nests followed in 1972 indicated an essentially normal curve that peaked in early June and then tapered off almost as rapidly as it had initially increased. Gray (1980) showed estimated timing of nest initiation for two years, of which one (1976) showed a gradual curve peaking in mid-June and a possible minor second peak in early July. The other showed a more irregular pattern, with laying peaks in mid-June and again in mid-July. Gray suggested that late initial nesting or, more probably, renesting attempts might be responsible for these results. The approximate four-month length of the nesting and hatching cycle in Gray's study area is seemingly too short to favor double-brooding, but renesting following initial nesting failures may be frequent.

Renesting by ruddy ducks has since been documented for Manitoba (Tome 1987) and may possibly contribute significantly to overall brood production in areas such as northern California (Rienecker and Anderson 1960), although it probably doesn't generally contribute much toward the species' overall reproductive success in North America (Bellrose 1980). Hughes (pers. com.) suggests that a potential seven-month breeding season (March to September) for ruddy ducks in Great Britain allows plenty of time for possible double-brooding to occur there.

ANNUAL RECRUITMENT AND POPULATION DYNAMICS

Estimates of reproductive success in any bird species are often quite difficult to obtain under natural conditions, for a variety of reasons. Among stiff-tailed ducks it is particularly difficult to make such estimates. This results in large part from the problems of estimating normal clutch size, which is often confused because of the effects of social parasitism and dump-nesting tendencies, and normal brood size, which is difficult to measure accurately because of early brood breakup as well as brood-merging tendencies. Joyner (1975) noted that in his study area the average brood size at hatching was 8.5 ducklings for 54 nests. Average brood size was reduced to 5.08 for 83 broods after the first two weeks. Thereafter brood reductions were minimal, but average brood sizes soon became impossible to judge, as broods tend to gather into flocks of 30 to 100 ducklings, with several females intermingling among them. For example, when 12 apparent broods were first observed by Joyner on one lake, they included 77 ducklings of two to four weeks in age. When these were then disturbed, they separated into apparent broods that included one female with 11 ducklings, two with 9 ducklings each, three with 7 each, two with 6 each, three with 4, and one with 3. After the groups had again merged, a second disturbance produced one "brood" of 11, one of 10, one of 9, one of 8, two of 7, two of 6, one of 5, and one (two?) of 4. Joyner noted that brood mergers most often occurred among broods of similar ages. Females invariably gaped at strange ducklings that were older and larger than their own, which helped prevent brood mergers of unequally aged ducklings. Brood mergers have also been observed among Australian blue-billed ducks, where an estimate of three surviving young per clutch has been made (Marchant and Higgins 1990).

The only stifftail species for which anything like a reasonable estimate of overall annual reproductive success is at all possible is the North American ruddy duck. Various components of annual reproductive success (clutch size, nesting success, hatching success, and mean brood sizes) are summarized in Table 19, using the result of a variety of field studies performed in several different parts of this species' breeding range. It is generally assumed that female North American ruddy ducks normally begin breeding at one year, and that most if not all of them attempt to nest every year. However, 4 of 19 females that Joyner (1975) collected during the prenesting period in Utah were possible nonbreeders, and Keith (1961) located only 0.73 nest per pair of resident ruddy ducks in his Alberta study area.

It seems reasonable to assume that second ruddy duck broods are produced only very occasionally in temperate North America, although the considerably longer breeding seasons occurring in the West Indies (December to May) and in South America (November to April) may make second broods more feasible there (Palmer 1976). If, indeed, second North American ruddy duck broods ever occur under natural conditions, they would be almost unique among temperate-breeding ducks of the Northern Hemisphere.

If we can further assume (from Table 19) that about two-thirds of the nesting female ruddy ducks succeed in hatching a brood and that brood size at hatching averages about 6 ducklings, then the average production of hatched young per nesting female would be 4.0 ducklings. There is likely to be a significant prefledging loss of young, especially in the first two weeks following hatching. Bellrose's (1980) estimates of duckling survival are about 78 percent (average reduction from 5.69 to 4.43 young per brood). This amount of duckling mortality seems reasonable, but must be regarded as only a guess based on the problems of accurate brood-size estimates mentioned earlier. Accepting this figure would suggest that each nesting female produces an average of about 3 fledged young. This means than 1.5 young should be recruited into the fall ruddy duck population per adult pair, assuming for simplicity's sake that a nearly equal adult sex ratio exists. This would in turn imply that the fall population should consist of approximately 60 percent immatures and that annual combined immature and adult mortality rates should also approximate 60 percent. Assuming a more realistic adult sex ratio containing 60 percent males, the percentage of immatures in the overall fall population should be about 50 percent.

One independent estimate of annual recruitment rates lies in the percentage of immature birds in hunter-kill samples. Such data represent an admittedly biased sampling method, since immatures are usually somewhat more vulnerable to death by hunting than are adults. Nevertheless, the overall mean adult-to-immature age ratio in hunter-kill samples of North American ruddy ducks from 1966 to 1973 was 1:1.72 (Bellrose 1980). This ratio represents a fall population composed of 63 percent immatures, which is very close to the estimate of recruitment rate based on the just-mentioned production figures that assumed equal adult sex ratios. However, Bellrose judged that immature ruddy ducks are probably significantly more prone to hunting mortality than are adults and believed that a fall age ratio of 1:1 (50 percent immatures) is more representative of the actual field situation. This would suggest that the actual ruddy duck recruitment rate (and its approximate annual first-year plus adult mortality rate) may indeed be fairly close to 50 percent, which is in agreement with the earlier calculations from estimates based on a mean female productivity of three fledged ducklings and adjusting for an unbalanced (60:40) adult male-to-female sex ratio.

Judging from relatively abundant available banding recovery data, annual mortality rates of several commonly hunted species of North American ducks—including mallards, North American black ducks, and blue-winged teal—average from about 40 to 60 percent (Farner 1955; Bellrose 1980). However, no stiff-tailed duck mortality rates have been been calculated by this method, because too few recoveries are yet available for any stifftail species to provide a basis for such an analysis.

If it is therefore realistic to believe that a female stifftail such as the North American ruddy duck or Australian blue-billed duck must on average produce three fledged young per year in order to maintain the population, then the adaptiveness of parasitic nesting in such "facultatively parasitic" species of *Oxyura* must be reconsidered. If the hatching success of parasitically laid eggs is no more than 20 to 25 percent, and if there is an additional minimum 25 percent posthatching but prefledging mortality of ducklings, then at least 15 eggs would have to be laid parasitically by an otherwise nonnesting female ruddy duck to make the equation "work" from a population-maintenance standpoint. This certainly seems unlikely and makes one believe, as did Joyner, that female ruddy ducks are not so much facultatively parasitic nesters as they are chance depositors of eggs in the nests of other birds, especially when host females have not yet completed the construction of their own nests.

Male North American ruddy duck.

PART TWO

Species Accounts

Pair of black-headed ducks swimming.

C H A P T E R   S I X

Black-Headed Duck

HETERONETTA ATRICAPILLA
Merrem 1841

Other Vernacular Names
    None in general English use; Kuckucksente, Schwarzkopfente
    (German); canard à tête noire de l'Argentine (French); pato
    sapo, pato riconero, pato de color pardo (Spanish); pato cabeza
    negra (Argentina and Uruguay).
Range of Species (Fig. 13)
    Resident in central Chile from Santiago to Valdivia, but appar-
    ently most abundant in central Argentina, from the lower east-
    ern slopes of the Andes to Buenos Aires. Probably also breeds
    in Uruguay, although specific records appear to be lacking.
    Also reported from but apparently rare in southeastern Bo-
    livia, Paraguay, and southeastern Brazil, with no known breed-
    ing records from these regions. Presumably mostly resident,
    but some northward and northwestward movement evidently
    occurs during winter in southern parts of the range.
Measurements, Adults and Immatures (Millimeters)
    Wing: males 157.5–178 (average of 71, 168.4), females 154–182
    (average of 62, 172.1). **Culmen:** males 40.7–47.0 (average of 67,
    44.1), females 41.0–48.1 (average of 57, 45.0). **Tail:** males 44–57
    (average of 70, 48.7), females 44–59 (average of 62, 52.1). **Tarsus:**

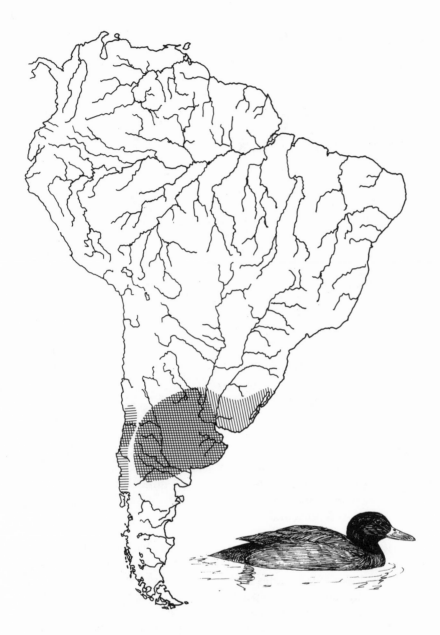

Figure 13. Breeding or residential distribution of the black-headed duck shown hatched; areas of probable denser populations shown cross-hatched.

males 30–34.5 (average of 71, 32.0), females 28–35 (average of 63, 32.25) (Weller 1967c).

Weights (Grams)

Eleven adult males 434–580 (average 512.6), 13 adult females 470–630 (average 565.2), representing an average male-to-female mass ratio of 0.9:1. Two immature males 350 and 370, three immature females 400–500 (average 453.3) (Weller 1967c). Two captive males averaged 400 (Carbonell 1983). One male 460, one female 605 (Belton 1984).

DESCRIPTION

Natal Plumage

The natal plumage is generally blackish-brown above and dark yellow below, but the facial area is tinted with orange-brown, with a dark facial stripe below the eye from the bill to the cheeks and a pale superciliary stripe, which is separated into two parts by a dark line from the eye to the crown, the anterior part being paler than the posterior part. Two dark yellow patches appear on the back, one on the pelvic area and the other on the pectoral area, forming a nearly continuous elongated band; a very small patch is also present on each wing. The upper mandible is nearly black, with yellow edges; the lower one, dull yellowish. Legs and feet are deep gray (Weller 1967c). A live duckling has been illustrated in color by Todd (1979).

Juvenal Plumage

Similar to that of the adult female, juvenal plumage is dark brown above and lighter below, with dark brown crown and eye stripes and a dark brown cross-barring on the feathers of the back, rump, and flanks. The abdomen is unmarked tannish-white. Undertail coverts are white with brown barring (females) to plain rufous (males). Wings have tan or dull white feather edgings; rectrices are as in adults, but the rachis are broad, untapered, and blunt-tipped (Weller, 1967c).

Immature (Basic I) Plumage

This plumage is very similar to the first nuptial plumage of females. There is more contrast than in the juvenal plumage, with a white abdomen and darker brown upperparts, the coarse, buffy cross-barring of the feathers being replaced by finer buff speckling at the feather edges.

Definitive Male Plumages (Breeding and Nonbreeding)

There is a complete molt in midsummer from the breeding to the nonbreeding (basic) plumage, and a partial molt in spring back into the breeding (alternate) plumage, but these are not certainly distinguishable in color from one another. The whole of the head and the upper neck are black, except for the chin, which is usually white. Entire upperparts are black, minutely freckled with rufous. The upper breast is brownish; the rest of the underparts are silvery white, with brown showing though. The sides and flanks are tawny to rufous, with black vermiculations. Undertail coverts are more brownish-red and are unmarked. A speculum is absent, but the secondaries have anterior and posterior whitish wing bars. The primaries and rectrices are blackish; most underwing coverts and axillars are white, the dorsal wing coverts gray. The bill is black dorsally, grading to grayish-blue toward the edges of the mandible, with flesh to yellowish-pink mandible edges. The lower mandible is often spotted with pale purple. The lateral basal portion of bill varies from bright rose (during breeding) to pale rose-pink (nonbreeding). The iris is dark brown, the legs and feet light green, with varied tones of gray, light brown, or flesh and grayish-brown webs (Weller 1967c).

Definitive Female Plumages (Breeding and Nonbreeding)

Two molts probably occur annually, but, as in males, no apparent seasonal variations exist in color patterns between the breeding and nonbreeding plumages. The body plumage is generally similar to that of the male, but the head is distinct, with the crown more brownish and the checks freckled with black and buff. A pale stripe extends from the bill to the eye, where it is interrupted, and continues again behind the eye as a gradually fading postocular stripe. The chin and throat are a dirty white. The iris is brown, the bill dark brown, with a tinge of yellowish-orange to yellowish-pink basally (duller in the nonbreeding season); legs and feet are pale slate, with darker webs.

IDENTIFICATION (FIG. 31B)

In the Hand

In contrast to the true stifftails, this species has a relatively short tail that is about the same length as the bill and is not easily distinguished from the adjoining coverts. The hind/toes lack the strong lobing typical of all other stifftails. The bill is narrower (under 20 millimeters) than any stifftail's except for the masked duck, and its nail is neither narrowed above nor recurved below.

In the Field

These birds tend to resemble surface-feeding ducks more than stifftails in the field, since they usually forage while swimming at the surface or tipping up in shallow water. However, males have notably thickened necks, black heads, and, unlike any other dabbling ducks of the region, bluish bills with bright pink to red basal spots. Females are superficially remarkably similar to females of various teals, such as the cinnamon teal (*Anas cyanoptera*), in their general plumage pattern, especially their head patterning, but both sexes lack the iridescent coloration on the wings typical of all South American dabbling ducks. They also do not "quack" like typical female dabbling ducks, and the tail is usually held low, at or near the water surface, rather than well above it as is typical of dabbling ducks. They take off readily and fly swiftly, with rapid wing beats, their heads held low, directly in line with their bodies rather than above body level.

ECOLOGY

Habitats and Densities

During the breeding season this species tends to occur in pairs in small pools of dense marshes, but later in the summer it concentrates on lakes and deep roadside ditches (Weller 1967b). It generally favors freshwater to slightly alkaline marshes that can be considered semipermanent to permanent and have extensive stands of emergent bulrushes *(Scirpus californicus)* and small-leaved floating vegetation such as duckweeds. In Argentina such marshes occur naturally along streams or are formed artificially where water is stored for irrigation and livestock. The birds are sometimes also found in wooded marshes where small patches of bulrushes occur and are commonly seen on flooded fields or even artificial lakes. They evidently are more influenced by the nature of the emergent vegetation than by the surrounding land habitats (Weller 1967a). Besides being a source of food and cover, the bulrushes are also bent over and used as submerged roost sites.

No information on densities is available.

Foods and Foraging

Based on the study of 27 specimens, including 22 adults and 5 full-grown immature birds, Weller (1968a) found that seeds of bulrushes composed this species' primary food. Seeds were found in 24 of the 27 birds and made up almost the entire food volume of 20 birds. (See Table 8.) During summer, the birds also eat snails (found in five specimens), which are swallowed whole. They also eat small seeds of other plants, including cutgrass *(Zizaniopsis)* and duckweed (Lemnaceae), especially the smaller varieties such as *Wolfiella* and *Lemna*.

When surface-feeding in duckweed, the birds swim with the emersed bill almost horizontal (Fig. 14D), but when they are straining mud in shallow water the bill is usually held at a diagonal angle. In slightly deeper water the birds tip up (Fig. 7E); when the water is too deep for tipping up they may dive for food in water up to a meter or so deep. When diving, the birds jump rather high before submerging (Fig. 14G). Dive durations were reported in Table 6. Several individuals observed by Weller (1968a) dived continuously for 45 to 55 minutes, often in company with coots and Argentine blue-billed ducks.

Competitors, Predators, and Symbionts

In the interspecific social relationships observed by Weller (1968a), all species of grebes, coots, and ducks other than the silver teal *(Anas versicolor)* clearly dominated black-headed ducks. Probably these latter consume some of the same foods as coots and various diving ducks, but it is unlikely that food limitations are ever a serious problem in their preferred marshes.

Predators known to inhabit the marshes used by black-headed ducks include two harriers *(Circus buffoni* and *C. cinereus),* which may take some young birds or perhaps even weak or disabled adults. The chimango and crested caracaras *(Milvago chimango* and *Polyborus plancus)* are also common in these marshes (Weller 1967c) and probably take some untended eggs. Caracaras are evidently not a direct threat to adults, and black-headed duck eggs have even been found in the nests of chimangos!

Important symbiont species are those that are sometimes used as hosts for eggs. Weller (1968a) listed at least 14 species that have been reported in the literature as host species and added one additional species to this list, the red-fronted coot *(Fulica rufifrons).* In his study areas of eastern Buenos Aires Province, the highest incidence of parasitism involved this species (69 percent of 133 nests) and the rosy-billed pochard *(Netta peposaca)* (83 percent of 6 nests). Other parasitized species were the red-gartered coot *(F. armillata)* (16 percent of 51 nests), the brown-hooded gull *(Larus maculipennis)* (14 percent of 7 nests), and the white-faced ibis *(Plegadis chihi)* (1.5 percent of 2071 nests). Success of some of these nests was tabulated earlier (Table 17) and will also be discussed below. (See Reproductive Success and Status).

ANNUAL CYCLE

Movements and Migrations

There are records of the species occurring as far north as Buena Vista, Sara Province, Bolivia, and other records for Paraguay and extreme southern Brazil, but there are no known breeding records for these countries. It is thus likely that these peripheral records are of vagrants, as are also perhaps some records for isolated marshes in the dry pampas of central Argentina and the dry foothills of the Andes. In the area around Buenos Aires, the birds are less common in fall and early winter than during spring and summer, and flocks of up of the 40 birds have been observed during winter in Sante Fe Province. As many as several thousand birds have been observed in Santiago del Estero Province during May. There thus seems to be a northward and northwestward movement from Argentine breeding areas following the breeding season (Weller 1967b).

Molts and Plumages

The best information on molts and plumages comes from the work of Weller (1967c), as has been summarized in the species' description given above. He found that a femalelike "first nonnuptial" (that is, first-winter) plumage is typical of this species but is unlike that of most Northern Hemisphere Anatidae. Weller also noted the unusually brief duration of the species' juvenal plumage and the presence of a long-lasting, nonbreeding (basic) adult plumage, which does not differ in color from the breeding (alternate) plumage. However, he found no evidence of two wing molts per year, which is in contrast to the situation apparently typical of all other stifftails. Instead, he observed that the remiges are molted once a year, during December and January, at the time the breeding plumage is being lost. Tail molt was observed in immatures collected by Weller during mid-January and in adult specimens obtained during early March. The tail molt is gradual and irregular, rather than nearly simultaneous as it is in at least some *Oxyura*.

Carbonell (1983) observed that the adult plumage appears in captive males at the Wildfowl Trust during their first year of life. In England there is a flightless period at the end of the breeding season (August to September), when the tail is also molted, and a second tail-molting period occurs in January before the breeding season. The remiges take about two weeks to regrow.

Breeding Cycle

Weller (1968a) observed in eastern Argentina that the breeding (egg-laying) season lasted about three months, from mid-September to mid-December, with a peak in October. This is slightly beyond the egg-laying curves Weller showed for the two coot species parasitized but apparently in approximate synchrony with the rosy-billed pochard's season. In England, the egg-laying period is also about three months (mid-April to mid-July), when nearly all other waterfowl are also laying (Carbonell 1983).

SOCIAL AND SEXUAL BEHAVIOR

Mating System

Weller (1968a) concluded from his field observations that pair bonds are formed and tested in a manner similar to those of other non-parasitic waterfowl species. He did say that some pairs appeared to be close-knit, whereas others even temporarily switched partners. Additionally, he found no evidence of permanent pair bonding; only 5 percent of the birds were seen in pairs during the post nesting period, as compared with 60 percent in pairs among those observed during the breeding period from early October through December. At the Wildfowl Trust pair bonding seemed to last for periods of as short as a few weeks to as long as a full year (Carbonell 1983). However, a great deal of seemingly promiscuous sexual activity occurred throughout the three breeding seasons observed by Carbonell and involved attempted copulations not only by unpaired males but also seemingly paired males. This would suggest that pair bonding is rather weak at best in the black-headed duck. However, Rees and Hillgarth (1984) observed that two males (of five observed at the Wildfowl Trust) had well-established pair bonds with females, feeding together much of the time, and they observed mutual preening of the neck feathers (a very rare form of social behavior in ducks) by paired birds on two occasions. The two most aggressive males were unpaired; both they and the paired males chased females a good deal of the time they were in view (up to 24 percent in the case of the most aggressive male).

Territoriality

There is no definite evidence of territoriality in this species, either in the wild or among captives at the Wildfowl Trust. Given the apparently weak pair bonding, and the fact that resources adequate for nesting and rearing of broods needn't be defended, there is no reason to suspect that territoriality should be present at all in this species, except perhaps as a possible female-attraction mechanism.

Courtship and Pair-Bonding Behavior

The first detailed account of the sexual displays of this species was that of Weller (1968a), which was based on his extended observations of wild birds. Carbonell's (1983) later observations of captives provides for a more detailed analysis and description of male displays. She observed that the normal length of male display bouts (4 to 5 males) was about 4 minutes; the maximum observed time was 17 minutes. Male displays observed by Carbonell were the toad call (55.4 percent of all filmed display occurrences), the head shake (44.6 percent of filmed occurrences), and the wing shake (none filmed). She did not observe turning the back of the head as had been described for a single captive male observed by Johnsgard (1965a). It now appears likely that this widespread pair-bonding mechanism of ducks is not an actual ritualized signal in this species, as it was not observed by Johnsgard in his more recent and much more extensive observations nor by others. Similarly, inciting by females is altogether lacking (Rees and Hillgarth 1984), which is another reason to question the strength of pair bonding in black-headed ducks.

The toad call ("gulping" of Johnsgard 1965a) (Fig. 14A). This is the primary male display of the species. It is performed either in front of or beside the female; at times a male will display when no females are present. A paired male will also perform it when another male approaches his female, while placing himself between the intruder and the female. The display has already been described by Johnsgard (1965a) and Weller (1968a). The male inflates his neck, lifts his tail and wings (tail-up, wing-up), and then lowers the tail while raising the head. While the wings and head are held at the extreme posture ("wing-lift-one" of Weller), the bill is slightly withdrawn in a pumplike manner and the wings are lowered. The head is then lowered and pulled backwards rapidly, and the wings are slightly raised again ("wing-lift-two" of Weller). Finally, shaking his tail, the male again assumes a normal swimming posture. A weak, whistlelike call is uttered during the head pumping, which Johnsgard described as a "pic-pic-pic." The average duration of 25 displays was 0.7 second (range 0.6–1.0, s.d. 0.11).

The head shake. This is a single lateral movement of the head, with

Figure 14. Social-sexual behavior of the black-headed duck: (A) toad-call sequence, (B) detail of toad-call posture, (C) head pumping, (D) female surface-feeding, (E) female threat posture, (F) wing-up, tail-up alarm posture, and (G) foraging dive posture. [Variously based on sketches from cine sequences by Carbonell (A, C), sketches or photos by Weller 1968a (E–G), and photos by Johnsgard (B, D).]

the bill taken slowly to one side and then rapidly moved to the opposite side before being returned forward. Of the total toad calls observed, 59.8 percent were preceded by a single head shake, 9.8 percent by two or three shakes, and 30.4 percent by none. The average duration of eight displays was 0.2 second (range 0.1–0.3, s.d. 0.05).

The wing shake. This movement was observed twice by Carbonell, once in a sequence of general toad calls, and once in a sequence of toad calls prior to copulation. It appears to be a ritualized form of comfort wing shaking.

Fertilization Behavior

Carbonell (1983) observed one full copulation sequence. The female was preening when the male began to swim toward her from several meters away. As he approached, he performed three toad calls, each preceded by a head shake, and one wing shake. The female then stopped preening, shook her tail, and began head-pumping movements identical to those of *Anas*. The male then performed several more toad call-head shakes and joined the female in a mutual head pumping. The female then lowered her body and the male mounted. After copulation, which lasted three to four seconds, the female began to preen again while the male performed two very rapid toad calls before he also started preening. The average duration of 13 male head pumps (Fig. 14C) was 0.24 second (range 0.1–0.3 second).

NESTING AND PARENTAL BEHAVIOR

Nest Choice and Egg Laying

Nesting choice by this parasitic species has been discussed in Chap. 5 and also previously under Competitors, Predators, and Symbionts. It appears that the species actually exhibits very little selective "choice" in depositing its eggs but rather places them in almost any nest encountered during the long egg-laying period. Observations by Michael Lubbock (pers. comm.) at the Wildfowl Trust in 1977 indicated that two eggs are normally laid in each host's nests, the eggs being deposited at daily intervals at about the time the host's clutch is complete.

Rees and Hillgarth (1984) reported that adults spend a good deal of time searching for nests, usually patrolling or skulking about in pairs, both during evening and early morning hours, but especially in the evening hours. Twice, when the female went into a nest box, the mate remained nearby, looking in occasionally. However, no eggs were laid.

One of the few first-hand account of actual egg laying in this species was provided by Powell (1979). A female at the Wildfowl Trust was observed to lay an egg within an approximate eight-minute period in the nest of the rosy-billed pochard during an early May evening. In a ten-day period between May 9, the date of this observation, and May 18 a total of seven eggs were laid, including three in the pochard nest, two in the nest of a common moorhen *(Gallinula chloropus)* and two in the nest of a cinnamon teal. Rees and Hillgarth (1984) observed one case of parasitism of a rosy-billed pochard's nest at the Wildfowl Trust. The female entered the nest while the female pochard was feeding and sat on it for eight minutes. At that time the male approached the nest. The pair then left, followed by the pair of pochards. At that time the pochard nest contained four eggs. Todd (1979) reported that Micheal Lubbock observed a female black-headed duck force an Argentine red shoveler off its nest and then laid an egg of its own within five minutes. Carbonell (1983) judged that two females at the Wildfowl Trust laid clusters of 3, 4, and 8 eggs, or an average of 5.0 per egg-laying series.

In Weller's (1968a) field study he found the distribution of eggs in 82 nests to be as follows: one egg in 73 percent of the nests two eggs in 21 percent, three eggs in 4 percent, five eggs in 1 percent, and eight eggs in 1 percent. Of course, some of these nests may have been parasitized by more than one female black-headed duck. Nores and Yzurieta (1980) reported that one or two eggs per nest are typical, but that up to six may be present, and Peña (1976) reported up to eight per nest.

Hatching and Brood-Related Behavior

Weller (1968a) estimated an incubation period under natural conditions of 24 to 25 days, which corresponds to Carbonell's (1983) determination of a 22 to 25 days, under a bantam hen or in artificial incubators. Todd (1979) believed that the black-headed ducklings are able to synchronize their time of hatching to conform with that of the host species, although this remains conjectural. Weller (1968a) observed that wild ducklings evidently remain in the host nest for only one or two days, that they tend to soon leave the broods with which they are hatched. This tendency develops as soon as the parasitic duckling is dry and mobile. They begin to strain small crustaceans from duckweed soon after hatching and thus appear to be fairly independent in a very short time. Fledging occurred at approximately 10 weeks for birds reared at the Wildfowl Trust. Maturity for one male was reached during his first year.

Additional observations on behavior of newly hatched ducklings at the Wildfowl Trust has been provided by Rees and Hillgarth (1984). Two eggs placed by these researchers in the nest of a rosy-billed pochard subsequently hatched. One of these hatched a day before any of the host's own clutch. It received no parental care from the host female and was rescued for fear that it might perish. The second one hatched with the pochard's clutch, and the female pochard treated this duckling like one of her own. She frequently waited for it, called to encourage it to follow, and on one occasion returned to collect it when it was tardy in joining her on the pond. The black-headed duckling soon became increasingly independent and was last seen with the brood four days after hatching. The next day it was caught and returned the propagation center for hand rearing. No diving was observed in these ducklings by the authors, who believed that this might be either a reason for considering *Heteronetta* as a dabbling duck or it might reflect the marshy habitat to which the species seems best adapted.

REPRODUCTIVE SUCCESS AND STATUS

Information on the nesting and hatching success of black-headed duck eggs under natural and captive conditions was summarized earlier in Chap. 5 (Tables 17 and 18). These data, although relatively limited, suggest a generally low level of reproductive success for the species under natural conditions, as compared with nonparasitic stiff-tails. Nevertheless, this reproductive strategy is clearly an evolved tactic and over the long run must have been more successful on average than nonparasitic nesting for it to have displaced such nesting behavior.

No general estimates of population sizes are available for black-headed ducks. The birds are certainly locally common but are dependent on appropriate and probably increasingly rare wetland habitats (permanent to semipermanent marshes with extensive emergent vegetation) and on the presence of an adequate host population of ducks, coots, and other suitable marsh-nesting species. Concentrations of nearly 250 birds have been seen in the coastal lagoons of Uruguay, and the species has also been reported in the Pantanal, an important wetland and waterfowl breeding area of southern Paraguay. The marshes of Cordoba Province, Argentina, are evidently important to this species; they have been observed in recent years on the Bañados de Rio Saldillo, the Lagunas de Etruria and Ludueña, and the Cañada de Los Tres Arboles and Los Morteros (Scott and Carbonell 1986).

Trio of masked ducks in social display.

CHAPTER SEVEN

Masked Duck

OXYURA DOMINICA
(L.) 1766

Other Vernacular Names
Black-masked duck, Dominican duck, quail duck, St. Domingo teal, spinous-tailed duck, squat duck, white-winged lake duck; Masken Ente, Maskenruderente (German); canard masqué, canard zombi (French West Indies); sarcelle á queue epíneuse (French); patico enmascarado, pato charretero, mano de piedra, pescuezo gordo (Venezuela); pato agostero, pato careto, pato chico, pato chorizo, pato codorniz (Spanish West Indies); pato colorado, pato criollo, pato dominico, pato enmascarado (Costa Rica, Mexico); pato espinoso, pato fierro (Argentina, Uruguay); pato sierro, pato tigre (Panama); pato zambullidor chico (Colombia); cau-cau, marrequinha, marreco-ca-ca (Brazil).

Range of Species (Fig. 15)
Local resident from coastal southern Texas (occasionally) through Mexico (scattered coastal records) and Costa Rica to Panama. In mainland South America in coastal and tropical freshwater wetlands from the interior and Pacific coast of Colombia south to northwestern Peru; also eastward through Venezuela, Surinam, Guyana, and probably much of the remaining eastern South American lowlands south to central Argentina (Cordoba, Santa Fe, and Buenos Aires provinces). Also

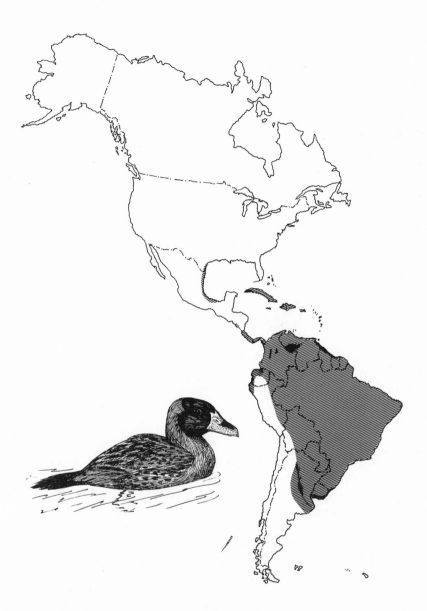

Figure 15. Breeding or residential distribution of the masked duck shown (hatched); areas of possible denser populations shown inked in.

widespread in the West Indies (especially the Greater Antilles of Cuba, Jamaica, and Hispaniola but has also nested in the Lesser Antilles on Martinique, St. Lucia, and perhaps Guadeloupe), and on Trinidad and Tobago. Probably residential throughout most of its range.

Measurements (Millimeters)

> **Wing:** males 136.2–138 (average of 5, 137.2), females 132–139 (average of 5, 136.4). **Culmen:** males 31.1–33.9 (average of 5, 32.0), females 30.8–34.2 (average of 5, 32.0). **Tail:** males 75.2–79 (average of 5, 77.1), females 74–79.2 (average of 4, 76.5). **Tarsus:** males 26.1–27.6 (average of 5, 27.0), females 25.7–27.9 (average of 5, 26.9) (Wetmore 1965).

Weights (Grams)

> Johnsgard (1975) reported that 4 males (3 adults and 1 immature) from various areas ranged from 386 to 410 (average 395) and 6 females (5 adult and 1 immature) ranged from 275 to 445 (average 338). Hartman (1961) reported 4 males averaging 359, and 3 females 368. Ripley and Watson (1956) noted an adult male at 386, a first-year male at 275, and 2 respectively aged females at 445 and 275. Haverschmidt (1972) reported that 9 males from Surinam ranged from 369 to 449 (average 406), and 6 females from 298 to 393 (average 339). The collective averages for these figures are 384.7 for 19 males and 346.2 for 17 females, representing a mean male-to-female mass ratio of 1.1:1. Gomez-Dallmeier and Cringan (1990) indicate an average weight for 29 adult males as 372, and 34 adult females as 358, or a mean mass ratio of 1.04:1. Wing length and other linear measurements shown above suggest that the adult mass ratio of the sexes should be virtually 1:1.

DESCRIPTION

Natal Down

Downy tail feathers are very well developed, being blackish above (sometimes more brownish, these possibly being older chicks or faded specimens) and yellowish-brown to buffy below (rather than whitish as in all other *Oxyura* species). The crown is sepia to fuscous, below which are two white or buffy superciliary and upper-cheek bands extending back from each eye toward the nape; these are separated by a distinct dark-brown eyestripe. A second distinct dark stripe on the cheeks is also present. [This stripe is much more distinct than as illustrated by Peter Scott's painting in Delacour (1959). A specimen illustrated by Bond (1961) is more indicative of the duckling's appearance than is Scott's painting in Delacour's book, which was evidently based on a badly faded specimen. A color painting of a downy duckling also appears in Gomez-Dallmeier and Cringan 1990.] Two pairs of very small yellow-buff patches appear on the sides of the back, one in the pectoral area near the wing, the other pair near the rump. There are no pale wing edgings. The iris is dark-brown; the bill is black or brown near the tip, with a paler nail, and

---

*Mostly after Palmer 1976, except where otherwise indicated.

orange basally; legs and feet are greenish-gray, with yellowish-white edging on the webs and toes.

Juvenal Plumage

Both sexes are superficially almost identical to the adult female, although with more uniformly colored underparts; the young bird is generally lighter and less blackish, its upperparts more heavily barred with cinnamon. The head has a broad, diffuse cheek stripe; the crown and upperparts are grayish-brown. Underparts varying from buffy to yellowish-brown (the latter possibly caused by staining). The tail is fuscous, nearly black; the feathers have protruding shafts that are blunt-tipped and often bare terminally, or sometimes down-tipped. The innermost secondaries have paler edges or whitish notches (Delacour 1959; Palmer 1976).

First-Winter (Basic I) Plumage

This plumage is not yet well described, but a highly contrasting face and body plumage pattern is definitely present. The rump is heavily barred with black and whitish, as are the sides and flanks. The abdomen is white, but the feathers around the vent are blotched with grayish. Birds in early stages of this plumage have juvenal tail feathers (Palmer 1976). "Subadults" have notably wider, paler margins to the feathers of the back and wing coverts than do adults, and more fluffy, almost downy feathers on the underparts, which produces a rather mottled visual effect (Ripley and Watson, 1956). Males are darker than adult females and have more pinkish-cinnamon on their underparts (Oberholser 1974).

Adult Breeding (Alternate) Plumage, Male

The definitive breeding plumage is presumably attained within one year. The head is mostly glossy black, changing abruptly to reddish-brown somewhat behind the eyes, this color extending down the neck and over the upperparts, breast, sides, and flanks. Elongated black spotting is present on most feathers of upperparts, varying in amount. Underparts and undertail coverts are whitish; the chin is also white. [Short (1976) has described an aberrant wild specimen from Paraguay that had a white facial patch and black and white speckling on the throat.] The tail is black, the wings mostly blackish above, except for a white speculum formed by the outer vanes of most (five to eight) secondaries and most or all the median and greater coverts. The innermost secondaries ("tertials") are blackish and somewhat tapered, the mostly white secondaries are broadly tipped with black, and the outermost secondaries are also blackish. Underwing coverts are deep brownish-gray, the axillars white. The iris is reddish-brown, with conspicuous narrow bluish eyelids; the bill is a vivid blue basally, paler distally, with a dorsal black stripe extending along culmen surface from nostrils to tip of nail; legs and feet are greenish-olive.

Adult Nonbreeding (Basic) Plumage, Male

The head is like the female in breeding plumage, but the lower facial stripe is usually wider, the chin is somewhat barred with black and whitish (not white). Upperpart feathers are notched laterally or have crossbars of buffy brown; the upper breast and upperparts are more washed-out in color; the underparts are generally whiter. The tail is very dark fuscous; the wing coverts are terminally margined and notched with tan; the innermost secondaries have broadly rounded tips. The iris is brown, the eyelids not noticeably bluish; the bill is mostly medium-grayish, becoming at least partly pale flesh-colored below; legs and feet are slaty.

Adult Breeding and Nonbreeding Plumages, Female

The sides of the head show two dark stripes; below the blackish crown is a light superciliary stripe, then a dark transocular stripe, next a broad light cheek stripe, then a dark cheek stripe, and finally a very pale buff to white on the lower cheek and chin. Dorsal feathers are patterned longitudinally darker, with reddish-brown margins (breeding) or cross-barred with buff (nonbreeding). The lower breast and abdomen are whitish (breeding) to pale cinnamon (nonbreeding). The tail is fuscous; the wing is between olive brown and fuscous, with a few cinnamon buff and russet spots. A broad white speculum is formed by the basal halves of most secondaries and the greater coverts, extending forward laterally to the outermost median coverts (this is slightly less extensive than in adult males). Underwing coverts are pale brown to grayish, the feathers have faint grayish-white edging and spotting; the axillars are white. The iris is brown; the bill slaty gray, tipped with a darker nail, and pink below. Legs and feet are grayish to brownish, much like the male's.

IDENTIFICATION (F. 31C)

In the Hand

The masked duck is easily distinguished from all other stifftails by its large white wing patch, which includes most of the secondaries and the adjoining coverts; additionally, the species' short bill (maximum length, 36 millimeters) is less than half the length of its relatively long tail (minimum length, 74 millimeters). Unlike that of all the other *Oxyura* species, the masked duck's outer toe is shorter than the middle toe, and the nail of the bill is not distinctly narrowed above.

In the Field

The masked duck is easily recognized in the field by its small size, its long tail (which is only rarely held above the water), and its rather short and relatively straight bill. Both sexes exhibit white wing patches, but these are not always visible in swimming birds. The species is prone to inhabit tropical marshes and swamps that are largely covered with water lilies and other floating vegetation, among which the birds are often effectively hidden from view. The birds take off rapidly but typically fly low over the water, usually disappearing quickly in the weeds as soon as they land or diving below the surface and hiding under surface vegetation when they finally again emerge. When taking off they may patter along the water for some distance. They can also take off directly, seemingly springing from the water like dabbling ducks, but probably first making a shallow dive and then propelling themselves out of the water in a rocketlike manner. In flight the birds resemble ruddy ducks, but their large white wing patches are clearly seen. In general they are extremely silent birds, like most stifftails. They have very large eyes and seem to be able to fly in nearly total darkness.

ECOLOGY

Habitats and Densities

Throughout its wide range the masked duck is associated with heavily overgrown swamps and marshes, where aquatic plants such as water lilies and water hyacinths are common. Mangrove swamps are sometimes also used, but apparently not for breeding. Extensively leaf-covered water surfaces, such as those favored by jacanas, are prime habitats for masked ducks, as the birds can effectively hide among these plants. In Venezuela the birds occur from sea level to about 200 meters along forested rivers and in marshes, in ponds heavily overgrown to vegetation, in rice fields, and occasionally along open sandy beaches (Gomez-Dallmeier and Cringan 1990). Wetmore (1918) stated that in Panama the birds inhabit freshwater ponds and the quiet waters of larger streams, where there are extensive growths of aquatic plants. From two to a dozen birds may sometimes be seen together, usually swimming in small pools or in open stands of floating vegetation.

There is no specific information on densities, but the birds are seemingly infrequent everywhere and evidently do not congregate significantly, either during the breeding or nonbreeding season. As many as 37 have been observed in Texas during winter (*Bull. Texas Orn. Soc.* 26:19, 1993).

Foods and Foraging

Little specific information is available. Phillips (1922-26) observed that foods from Cuban specimens mostly consist of the seeds of smartweeds (*Polygonum*), plus small amounts of water lily (*Castalia*), rush (*Fimbristylis*), dodder (*Cuscuta*) and "sawgrass" (*Cladium?*). Weller (1968a) noted several kinds of seeds in the gizzard of a single specimen that he obtained, and Dale Crider (pers. comm.) found wild millet (*Echinochloa*) seeds to be very important foods in northern Argentina. The shape of this species' bill suggests that it is better adapted for seed eating than is typical of most *Oxyura* but may be less well adapted for finding and extracting mud-bottom invertebrates such as midge larvae.

While foraging, the bird typically dives in fairly shallow water, and some data of diving and resting times for masked ducks during foraging was summarized earlier in Table 6. Additionally, Dirk Hagemeyer determined that an adult female remained submerged for 23 to 26 seconds during its foraging dives, taking intervening rest periods of 9 to 12 seconds, while diving in water about 2 meters deep. It foraged in this manner for about 45 minutes. At the same general location young birds were seen diving for periods of 15 to 17 seconds in water about a meter deep. These birds would forage for 15 to 30 minutes and then retire to a nearby shore to preen and rest for about 90 minutes (Johnsgard 1975).

Competitors, Predators, and Symbionts

There is little direct information in this regard for the masked duck. Although they certainly must feed on quite different materials, masked ducks are very commonly seen in company with least grebes (*Podiceps dominicus*) and common jacanas (*Jacana spinosa*) in Johnsgard's experiences in the West Indies and Colombia; this association has also been reported as typical in Venezuela by Gomez-Dallmeier and Cringan (1990). At times the species probably also coexists with ruddy ducks, but direct interactions between the two have not be reported. In Argentina they occur in habitats sometimes used by black-headed ducks as well as in habitats of rosy-billed pochards and fulvous whistling ducks (*Dendrocygna bicolor*).

Crested caracaras (*Caracara plancus*) are perhaps significant egg predators over much of South America, as is known to be true for at least most ducks in northern Argentina. There, eggs of black-headed ducks have frequently been seen in masked duck nests by Dale Crider (Johnsgard 1975).

ANNUAL CYCLE

Movements and Migrations

No detailed information is yet available. In Texas, at the northern extreme of their range, the birds have been observed both during summer and winter months, so seasonal movements there are probably not significant. The same may well be true elsewhere in the species' range, but this has not yet been documented.

Molts and Plumages

Molts of this species are still inadequately known. Palmer (1976) has attempted to describe the plumages (summarized above) from museum species, but much remains uncertain. Palmer noted that there is a gradual tail molt, the order of feather replacement being apparently random. He judged that there are two head, body, and tail molts per year, spaced about a half-year apart (probably during late fall and again in late spring or summer in Texas). He also concluded that much, perhaps all, of the wing feathering is molted twice per annual cycle but the timing of the wing molt in relation to general body molt is still undetermined. Dale Crider (pers. comm.) observed that a good deal of variation in the timing of the postbreeding molt was evident among wild birds in Argentina, since some males were flightless while others were still in full breeding plumage.

Breeding Cycle

Breeding seasons are highly variable throughout this species' range. Males in full breeding plumage have been obtained essentially throughout the year, although they compose a distinct minority of museum specimens. Breeding season is probably associated with the wet season, rather than summer as such, although in the tropics the summer tends to be the wetter period. Too few nests have yet been found in any single area to provide much information on the duration of the season. A few Texas nests have been found in September and October, but young have been observed between early May and mid-November (Oberholser 1974; Delnicki 1975). In Cuba nests have been found between early June and September (Johnsgard and Hagemeyer 1969), and in Jamaica nesting is said to occur from June to October (Downer and Sutton 1990). In Venezuela breeding occurs between April and September during the rainy season when water levels in rice fields and natural wetlands are increasing (Gomez-Dallmeier and Cringan 1990).

SOCIAL AND SEXUAL BEHAVIOR

Mating System

On the basis of limited observations of seemingly paired birds prior to breeding, Johnsgard (1975) judged that the masked duck is probably monogamous, with temporary pair bonds. However, broods that have been seen in Texas were always observed to be led by femalelike birds, suggesting that pair bonds are broken before hatching occurs.

Territoriality

No evidence yet exists that males exhibit territorial behavior. Their general elusiveness would suggest that any advertising behavior that they do perform must be quite limited or inconspicuous.

Courtship and Pair-Bonding Behavior

Less information exists on this topic than is the case for any other stifftail species, in spite of repeated attempts by both present authors to observe it among wild birds. Undescribed courtship behavior has been reportedly seen during April in Texas and May in Louisiana (Johnsgard 1975). Several relatively untrained observers have compared the male's display with the typical male display (bubbling sequence) of the North American ruddy duck, although this comparison now seems highly questionable and unlikely. One of the few

Figure 16. Social-sexual behavior of the masked duck: (A, B) golf-ball display of male and (C) chin lifting by female. (After photos by Rod Hall.)

useful recent observations comes from Rod Hall of the British Airways Assisting Conservation program. Near the shore of a mostly vegetation-covered lake in the Greater Antilles he initially observed a female sitting in open water, seeming to tread water. A few minutes later a male appeared and slowly approached the female with his head low and neck fairly large (Fig. 16A). He then expanded his upper neck to golf-ball size and uttered a steady "coo-coo-coo" similar to that of a domestic pigeon (Fig. 16B). The female silently faced the male with her bill somewhat raised and neck stretched upward (Fig. 16C). The male's tail was not cocked during this entire period of several minutes, and the female was apparently silent. At no time did the birds approach one another closer than about half a meter, and the female eventually turned and departed (Rod Hall, pers. comm.). Others have described what may well be the same call as a repeated and descending "du-du-du," a dull and almost inaudible "oo-oo-oo," or a repeated "kirroo-kirroo." Some early published descriptions of male display involving breast-beating behavior seem to have resulted from confusion with the displays of North American ruddy ducks.

Fertilization Behavior

There is no information available.

NESTING AND PARENTAL BEHAVIOR

Nests and Nest Building

The nests of masked ducks are usually very close to water. Bond (1961) described one as being a deep cup of rice stems placed just above water level. In Panama they have been found among rushes (Phillips 1922-26). In Argentina, Dale Crider observed that the nests were often placed in rice fields, amid rice clumps and beside deep water into which the incubating female could easily escape. These nests were described as being roofed over and basketballlike, with lateral entries. Only a small amount of down was present (Johnsgard 1975).

Egg Laying and Incubation

The laying rate is probably one egg per day. Egg sizes reported by Bond (1958) from Cuba and shown in Table 13 are somewhat smaller than an egg measuring 63 x 46 found in Panama (Wetmore 1965) or four from Trinidad, which averaged 60 x 46 (ffrench 1991). If the latter measurements are more typical, the estimated egg mass would be 70 grams, and newly hatched ducklings should weigh about 40 grams. Two "day-old" ducklings from Trinidad averaged 32 grams (ffrench, 1991), or very close to the predicted hatching weight of 28 grams, based on the smaller egg measurements shown in Table 14.

The average clutch size of masked ducks is generally believed to be of four to six (Wetmore 1965) or five to six eggs (Phillips 1922–26). One Texas nest contained six eggs (*Audubon Field Notes* 22:625, 1968). However, apparent dump-nesting or possible parasitic egg laying in areas of dense breeding concentration often increases the apparent clutch size, and as many as 27 eggs have been reported in a single Argentine nest. Bond (1961) mentioned eight Cuban clutches from Pinar del Rio Province containing 8 to 18 eggs (average 12.5), and Dale Crider observed that the usual clutches in Argentina were of about 10 eggs. In both of these areas breeding is apparently fairly common in rice plantations, and dump-nesting seems quite possible.

The incubation period is uncertain but has been estimated at about 28 days, which seems an unusually long period for such small eggs. Females certainly do all the incubation, but males may remain in the vicinity of the nest for some time into the incubation period (Johnsgard 1975).

Hatching and Brood-Related Behavior

There is little information on newly hatched brood behavior or brood sizes. One brood studied in Texas by Dirk Hagemeyer had four ducklings, which were simultaneously tended by two femalelike birds. The brood scarcely moved more than a hundred meters throughout the entire period of observation, which lasted for about 45 days, when fledging evidently occurred. Dale Crider noted that in Argentina the broods were often brought back to the nest for nighttime brooding, while molting of postbreeding or nonbreeding adults simultaneously occurred in natural ponds adjacent to the rice fields.

REPRODUCTIVE SUCCESS AND STATUS

No information on reproductive success exists. Probably the species is most common in rice-growing areas of Cuba and Argentina, according to the number of nests that have been found, but no quantitative information exists on that topic either. The species' status in the United States is marginal at best. There was an apparent influx of birds into Texas during the 1960s, and since then a few masked ducks have been seen almost every year in that state or in Florida. During the early 1960s there were a large number of U.S. reports of masked ducks, often occurring after the passage of Caribbean hurricanes. These included sightings in New Jersey, Georgia, Florida, Tennessee, and even Iowa (the last being a record later considered doubtful). Additionally, in Texas the birds were sighted on numerous occasions and in several southeastern and coastal counties (Cameron, Willacy, Kenedy, Brooks, Jim Wells, Nueces, Brazoria, Calhoun) during the 1960s.

During the 1970s and 1980s most U.S. sightings of masked ducks occurred in coastal Texas, with notable additional inland sightings in Dallas, Hays, and Bexar counties and at Big Bend National Park. Florida was second in frequency of masked duck occurrences, with many sightings at Loxahatchee National Wildlife Refuge, Palm Beach County. Additionally, there were new sightings in Louisiana (now with at least six records) and initial state records for Pennsylvania, Alabama, and North Carolina.

The first definite U.S. breeding since the 1930s occurred in the autumn of 1967, when a brood was observed in Chambers County, Texas. A nest was found the following year in Brooks County, Texas (Johnsgard and Hagemeyer 1969). Earlier, broods had been reported on three occasions during the 1930s in Cameron County, Texas (Oberholser 1974), but there have been no proven nestings since 1968 in

Texas or elsewhere in the United States, and there has been a seemingly general decline in occurrences.

Masked ducks have been reported on the Audubon Christmas Bird Counts on seven occasions since 1967. The majority of such sightings (five) have occurred in Texas localities (Welder Wildlife Refuge twice, Bolivar Peninsula, Santa Ana National Wildlife Refuge, and La Sal Vieja), but twice the birds were seen in Florida (West Palm Beach and Lake Wales). Masked ducks were observed on all the Audubon Christmas counts from 1967 to 1970 but have only been seen on three such counts since then. The largest number of individuals recorded at a single locality for the period 1967–91 was 13 during a Christmas count in 1968; the birds were seen at Welder Wildlife Refuge, San Patricio County, Texas. As many as four pairs were present there during June of 1993 (after a unusually wet spring), and one nest was found to be under construction that month. Four adults and seven young were seen in Jim Wells County that same month (*American Birds* 47:1126, 1993). There was apparently a major movement into Texas during 1993, when as many as 100 birds were believed to be present that winter. The overall pattern in Texas would seem to be a movement into the state after wet summers. The numbers typically peak during winter, and during some years a few birds remain to breed the following summer. Males gradually molt out of their femalelike plumages over the winter, but full-plumaged males would seem to make up a distinct minority, even during summer.

The distribution of masked ducks in South America is broad, but the population is seemingly rather thin everywhere. In recent years the birds have been observed in good numbers on the Bajos Submeridionales and in Chaco and Santa Fe provinces, Argentina and on pampas wetlands of that country's Cordoba Province, such as Laguna Ludueña and Laguna Santo Domingo (Scott and Carbonell 1986; Nores and Yzurieta 1980). They have also been found to be locally common in Tucuman Province, northwestern Argentina (Dale Crider, pers. comm.) and probably extend into the Paraguayan chaco. They are evidently widespread in Uruguay, at least in the south and along the coastal lagoons (Barattini and Escalante 1971; Scott and Carbonell 1986). In Brazil they have been reported in recent years from the lagoons of Rio de Janeiro, but there are surprisingly few locality records for that vast country. In Surinam they are probably mostly of coastal distribution; they were reported as "not uncommon" on the Mahaicony River in 1965 (Snyder 1966). The masked duck is also fairly common on various freshwater habitats of Surinam (Haverschmidt 1958, 1972). The species also occurs along the coastal lowlands of Venezuela, as well as in the extensive interior Venezuelan llanos wetlands of the Orinoco basin (Scott and Carbonell 1986; Gomez-Dallmeier and Cringan 1990).

In Central America the masked duck has been reported from numerous Panamanian localities, mainly on the Pacific slope and coast; it is perhaps most common on the Santa Maria marshes of Herrera Province. It also occurs locally in Darien and Veraguas provinces and on the Atlantic slope on the Chagres River (Wetmore 1965; Mendez 1979; Scott and Carbonell 1986). It also probably breeds locally in Costa Rica, such as in Guanacaste Province. However, farther to the north, in Nicaragua, Honduras, El Salvador, Guatemala, and Belize the species is distinctly rare, though up to 500 birds have been seen during winter on Jocotal Lagoon, El Salvador (Scott and Carbonell 1986). There are evidently few if any specific breeding records for Mexico, but local breeding may well occur in the coastal wetlands of Sinaloa and Nayarit on the Pacific Coast and probably also along the coastal lowlands of the Carribean coastline.

The relatively few recent records of the masked duck from the West Indies are largely associated with coastal estuarine lagoons or freshwater marshes. In Cuba many of the available nesting records are from Pinar del Rio Province and are associated with the rice culture. Presumably Cuba is the primary source of the masked ducks that periodically appear in Florida and elsewhere in the southeastern United States following tropical storms.

A recent world estimate of population size was of 10,000 individuals, (Callaghan and Green 1993), but this can only be a crude guess at best.

Courting party of North American ruddy ducks.

Ruddy Duck

OXYURA JAMAICENSIS
(Gmelin) 1789

Other Vernacular Names

For *jamaicensis*: North American ruddy duck; also West Indian or Antillean ruddy duck (if subspecifically distinguished from mainland form). Numerous hunter's or local vernacular names exist, including blue-bill, booby, bullneck, butterball, diving teal, hardhead, red diver, rubber duck, and spinetail; Amerikanische Ruderente (German); erismature rousse, erismature roux, erismature à joues blancs, le canard roux (French); malvasia cariblanca, pato espinosa, pato chorizo, pato floridano, pato rojo americano, pato zambullidor (Spanish).

For *andina*: Colombian ruddy duck; Andean ruddy duck, Colombian lake duck. See *ferruginea* for relevant non-English names.

For *ferruginea*: Peruvian ruddy duck. Also Andean duck, Andean lake duck, Andean ruddy duck, ferruginous duck, Peruvian lake duck; Peruvianische ruderente, Schwarzkopf ruderente (German); erismature à tête noire (French); pato malvasia, pato pana, pato pitroco, pato rana de pico ancho, pato tripoco (Spanish). In Argentina, pato zambullidor grande is used to distinguish this species from *vittata*, which is called pato zambullidor chico.

Ranges of Subspecies (Fig. 17)

Of nominate *jamaicensis,* including the questionably separable mainland North American race *rubida* (Wilson) 1814: Breeds from central British Columbia (occasionally to southern Alaska) and Great Slave Lake east to southern Manitoba and Minnesota, and south to northern Baja California, Arizona, New Mexico, Nebraska, and Iowa, with more scattered and sporadic breeding records east to New York and south to Florida, coastal Texas, and (locally and/or sporadically) interior central Mexico. Winters from Puget Sound south to Chiapas, especially along the coasts of California and northwestern Mexico south to about Guerrero; also from Massachusetts south to Florida, mainly from Chesapeake Bay to Pamlico Sound, and in the Gulf Coast, especially along the Louisiana and Texas coasts and adjacent coastal Tamaulipas. Resident locally in the West Indies, including Cuba, Jamaica, Hispaniola, Puerto Rico (common but local), Virgin Islands (very uncommon), and south apparently rather erratically through the Lesser Antilles to the Grenadines and Grenada. The bird is also self-introduced (as feral escapees from captivity in 1952) and now well established in England. Self-introduced since 1965 into mainland Europe, it has been the subject of recent breeding reports from the Netherlands, France, Belgium, and Spain; it is also reported from 10 other mainland European countries plus Iceland.

Of *andina* **Lehmann 1946:** Endemic to the highland lakes of the central and eastern Andes in Colombia, where it is apparently restricted to the lakes of the Savanna de Bogotá (Departments of Boyacá and Cundinamarca), the surrounding paramo lagoons in the alpine zone of the Cordillera Oriental, and the adjoining Cordillera Central (Departments of Risaraldas and Cauca). The southernmost specimen record for this race is apparently for Purace, Cauca Province (Hellmayr and Conover 1948). The validity of this race has been questioned by Todd (1979), who considers it as an intergrade population. Fjeldså (1986) believes it might represent a Pleistocene relict hybrid population involving *jamaicensis* and *ferruginea*, with most male individuals approaching one or the other of these two extreme phenotypes.

Of *ferruginea* **Eyton 1836:** Resident on highland lakes and marshes of equatorial Andean Ecuador (Lago Yaguarcocha, Imbabura Province; Lago Antisana, Pichincha Province; Lago San Pablo, Otavalo Province). Also disjunctively on lakes and marshes of the more Peruvian Andes (Lago Junin and Lagunas Chacaycancha, Cutaycocha, and Sallahu, Junin Department; Lago Titicaca and Lagunas Lagunillas and Sallahu, Puno Department; Lagunas de Marcapomacocha, Pasco Department; Lagunas Pomacanchi, Asnacocha, and Pampa Marca, Cuzco

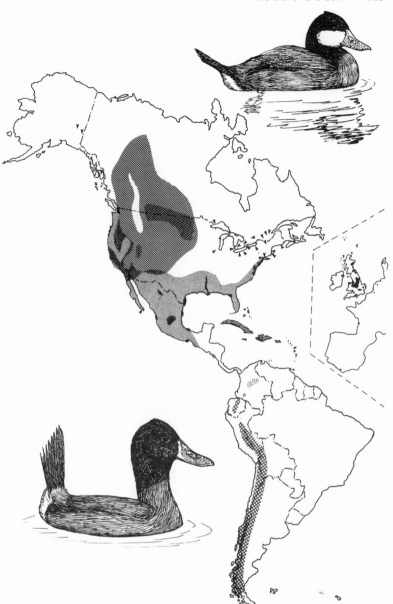

Figure 17. Native distribution of the ruddy duck, including the South American residential ranges of its Andean (cross-hatched) and Colombian (stippled) races. Extralimital nestings or minor breeding areas are shown by pointers. The residential West Indian and the summer breeding ranges of the North American race are hatched, with areas of denser populations indicated by cross-hatching. The North American race's wintering range is shown by fine stippling, with areas of major concentration shown in black. The introduced United Kingdom and European distribution is shown in the inset at right.

Department; and Lagunas de Yaurihuiri, Ayocucho Department) and in the Bolivian Andes (Uru-Uru, Oruro Department; Vacas, Corani, and Alalay, Cochabamba Department; Huaqui, La Paz Department). At progressively diminishing altitudes, south through the Chilean Andes (Parinacota, Tarapaca Province; Piuquenes, Coquimbo Province; Encañado; Santiago Province; Las Truchas, Ñuble Province; Laja, Bio-Bio Province) probably at least occasionally to Tierra del Fuego (Sarita and Gente Grande). Reportedly ranges north to the southern Colombian Andes, but a few specimens from this area (Purace, Cauca Province) are of *andina* (Hellmayr and Conover 1948). Also reported from sea level near Lima, Peru (Scott and Carbonell, 1986). Resident on low-altitude lakes and rivers of south-central Chile from Aconacagua south (Lago Peñuelas, Valparaiso Province; Villarrica and Lago Malleco, Cautin Province; Rauco, Valdivia Province; Rio Nirihuao; Aysen Province). Also resident on Andean lakes of Argentina (Lagos Argentino and Viedma, Santa Cruz Province) and temperate wetlands of the Argentine pampas (Laguna La Margarita, Cordoba Province) from Rio Negro southward to Tierra del Fuego (Viamonte). Considered a separate species by Livezey (1995).

Measurements (Millimeters)

Of *jamaicensis* **from continental North America** *Wing;* males 146–153 (average of 12, 150), females 139–150 (average of 12, 146). **Culmen:** males 36–44 (average of 12, 41.6), females 38–43 (average of 12, 40.3). **Tail:** males 68–78 (average of 12, 72), females 63–74 (average of 12, 70). **Tarsus:** males 32–35 (average of 12, 33.4), females 31–34 (average of 12, 33) (Palmer 1976). **Maximum bill widths:** males 21.2–24.8 (average of 13, 23.0) (Johnsgard pers. obs.).

Of *andina* (sample sizes not stated) **Wing:** males 139–149, females 133–143. **Culmen:** males 40—45, females 37–42. **Tail:** males 67–78. **Tarsus:** males 31–37. **Maximum bill width:** both sexes 24.3 (Delacour 1959), two males 25 and 29 (Lehmann 1946).

Of *ferruginea* (sample sizes not stated unless indicated) **Wing,** males 146–163, females 143–154. **Culmen:** males 40–45, females 40–45. **Tarsus:** males 36–38, females 34–36 (Delacour 1959). **Wing:** males 158–165, females 152–155 (Hellmayr and Conover 1948). **Maximum bill width:** 27 males 23–26.1 (Johnsgard 1965a); both sexes 24–26 (Phillips 1922–26). **Means of both sexes:** wing 162, culmen 45.2, tarsus 38.2, maximum bill width 26.9 (Johnson 1965).

Weights (Grams)

Of *jamaicensis* **in continental North America** Males 539–794 (average of 8, 610), females 310–650 (average of 13, 511) (Kortright 1943). Adult males in fall, 454–539 (average of 4, 482), adult females in fall, 454–595 (average of 3, 539) (Jahn and Hunt 1964). Twelve wild males averaged 590, and 17 females 499 (Palmer 1976). Captives in England, males 545–670 (average of 6, 610.8), females 500-580 (average of 7, 537.9) (Carbonell 1983). The collective mean averages for these four samples are 584.9 for 30 males versus 513.4 for 40 females, representing a mean male-to-female mass ratio of 1.1:1.

Of late-winter birds collected in California, 6 immature females averaged 506, 8 adult females averaged 529, 2 immature males averaged 510, and 4 adult males averaged 598 (Hohman, Ankney, and Roster 1992). Weights of 19 females during April-May (including 4 probable immatures) ranged from 394.5 to 629 (average 531.8). The 4 probable immatures ranged from 394 to 522 and averaged 474 (Joyner 1975). Winter (January and February) weights of 21 males in California averaged 516 grams versus 481 grams for 6 males in March. Average male weights increased in May and were fairly stable through summer, ranging between about 600 and 650 grams, and then declined to an average of about 575 grams during August and September. Mean weight of 4 females in winter was 491 grams. This was followed by a slight average weight decline after spring migration, but then by large increases in average weights during the prelaying and laying periods. Nine egg-laying females averaged 817 grams, with a maximum of 920. This peak weight was followed by an average decline of 125 grams during the approximate seven-day laying period. By the end of the incubation period the average weight of 4 females was just under 475 grams, an approximate 50 percent reduction from the maximum weights observed during egg laying (Gray 1980). Average spring arrival weights of 15 females in Manitoba averaged 582, which increased to 705 for 9 prelaying birds. Six laying females averaged 739, which declined to 594 early in incubation (9 females) and 509 late in incubation (8 females). The average of 5 brooding females was 546 (Tome 1984).

Of *andina.* No information.

Of *ferruginea.* One male, 822 (Humphrey et al. 1970). Males (sample size not stated) 817–848 (Todd 1979).

DESCRIPTION*

Natal Down

The down of *jamaicensis, andina,* and *ferruginea* are very similar (but progressively darker), being dark blackish-gray above and dirty white below. A whitish band extends from the bill to the nape below

Mostly after Palmer 1976 except as indicated.

the eyes (this band is lacking in *ferruginea*), and a dark brown streak white below. A whitish band extends from the bill to the nape below the eyes (this band is lacking in *ferruginea*), and a dark brown streak runs through the cheeks below. Chin and underparts are grayish-white; throat, neck, and sides of head are freckled with whitish (tinged with reddish in *ferruginea*); and chest band is a dark brown. Downy *jamaicensis* usually have two whitish scapular spots on the upper back and two smaller ones in the pelvic region; these spots are smaller in *andina* and nearly or entirely lacking in *ferruginea*. Downy tail feathers are dark brown and fanlike, with long, stiff shafts. The iris is very dark brown, the bill is grayish-black, the lower mandible somewhat pinkish to yellow, legs and feet black. The tracheal air sac of the male is present at hatching, at least in *jamaicensis*.

Juvenal Plumage (of jamaicensis)

The sexes are apparently alike and similar to the adult female, but the dark brown is grayer and duller. The crown is dull blackish-brown, with a dark cheek stripe. Upperparts are dull black to blackish-brown, with narrow crossbars. (These parts are less speckled and more barred than in adults.) The flanks are pale brown, with undertail coverts becoming white; both areas have a brownish barring; the abdomen is silvery gray. Feathers of the underparts and flanks are narrow, with grayish-white tips, giving a distinctive scaly appearance. Tail feathers are a dark brown and have blunt shaft tips. The innermost secondaries are short and straight; upperwing coverts are narrow and have somewhat pointed rather than rounded tips (Cramp and Simmons 1977). The bill is a dark-clove-brown on the upper mandible; the lower mandible is paler and mottled with pinkish; the legs and feet are clove-brown (males) or olive-gray (females) (Oberholser 1974). The juvenal plumage of *andina* is undescribed; that of *ferruginea* resembles *jamaicensis* but is generally darker; the black of the crown comes down to well below the eyes, the lores are blackish rather than light gray or buff, and the cheeks are more heavily streaked. The upper breast is a rich rusty brown with black barring, and there is no rusty or reddish tinge to the upperparts (Phillips 1922–26; Hughes, pers. com.).

First-Winter (Basic I) Plumage of jamaicensis

Both sexes are generally like the adult female, but tail and abdominal feathers are still of juvenal type, at least during early stages of this plumage. Palmer (1976) reported that this plumage results from a complete molt that includes acquiring new wing feathers, but Hughes (1990) observed that the juvenal wing feathers were retained when his captive birds underwent their molt into the first-winter plumage.

Male. The male's crown is a mixture of browns and blacks; its cheeks have a diffuse dark area or are even largely white; the throat is gray; the upperparts dusky, with pepperlike marks or irregular coarse bars; sides and underparts are silvery (Palmer 1976). The bill and feet are black; the legs brown; the iris brownish-black.

Female. The female's first-winter plumage is much like that of adult female, but rump and upperpart patterns are generally more barred and less speckled than in adults. Joyner (1975) observed that four apparently immature females he collected in April (one was flightless, three retained bursal remnants) had faded and highly worn rectrices and dull, buffy-speckled back feathers, without any reddish speckling on the scapulars and interscapulars, nor any adultlike reddish crown barring. The bill is dark clove-brown on the upper mandible; the lower mandible is mostly sepia; the iris is brown; the legs and feet bluish-olive gray, with the outer sides and joints black (Oberholser 1974).

First-Winter Plumages of andina and ferruginea

These plumages are undescribed in detail; Phillips (1922-26) stated that young males initially resemble females, but gradually acquire reddish feathers on the mantle or scapulars.

Adult Breeding (Alternate) Plumage, Males of jamaicensis

The breeding plumage is attained following a complete molt that includes the wing feathers (Hughes 1990), contrary to Palmer (1976), who believed that the first-winter wing feathers are retained. First-year males have white cheeks like older males, but have some dull-brown feathers mixed with the definitive chestnut body feathers and attain breeding condition later than do older males. On the head, the forehead, crown, and nape are black; the cheeks white Upperparts, sides, and flanks are tawny to deep reddish chestnut, sometimes fading to yellowish-brown. Underparts are mostly silvery, with darker feather bases sometimes showing though. The tail is fuscous brown. The wings are very dark fuscous brown above, the axillars and underwing coverts paler and mixed with white and brownish-gray. A varied amount of bright chestnut is present on the inner greater and median coverts. The innermost secondaries are elongated and sickle-shaped, their coverts having rounded tips. The iris is reddish-brown, the bill turquoise-cobalt, with a brownish nail, becoming pinkish at edges of mandible; legs and feet are bluish-gray, with darker webs. Males of *andina* are generally similar but show a great deal of individual variation in cheek spotting (Fig. 18). *Ferruginea* males usually have all-black heads, but some white flecking may be apparent on the cheeks, and a variably large white chin patch is typically present. A melanistic male individual of *jamaicensis* that resembled *ferruginea* was illustrated by Todd (1979), who also stated that second-generation male hybrids between *jamaicensis* and *ferruginea* closely resemble males of *andina* in their cheek spotting.

Figure 18. Left: Variations in head plumages of adult male Colombian ruddy ducks. (After Adams and Slavid 1984 and Fjeldså 1986.) Right: First-year male white-headed ducks. (After Amat and Sanchez 1982.)

Adult Nonbreeding (Basic) Plumage, Males of jamaicensis

The forehead and crown are very dark, becoming lighter toward the nape; the cheeks are white, most of the neck smoky-gray. Mantle and scapulars are dark-brownish with fine darker markings; the lower back is dusky brown. The upper breast is gray with a cinnamon tinge; sides and flanks are grayish, with brownish to chestnut mottling. The tail is fuscous, nearly black; the wing is fuscous as in breeding plumage but the innermost secondaries are shorter, nearly straight, and with more rounded tips. The upper mandible is black, the lower mandible raw umber (Oberholser 1974). Males of *ferruginea* are mostly dark brown, finely vermiculated with whitish on the upperparts; the crown, face, and upper throat are blackish; the lower throat and sides of head are finely freckled with white (Delacour 1959).

Adult Breeding Plumage, Females of jamaicensis

The forehead, crown, and nape are black and brownish mixed; cheeks are mostly white, with a dark facial stripe; upperparts are dark-brownish, speckled, and barred with light browns and sometimes tinged with chestnut; the upper breast is brownish; the lower breast to tail, silvery white. Tail feathers and remiges are fuscous; upperwing coverts are dull brown, some secondaries are narrowly tipped with white, and the innermost secondaries are long, curved, and tapering. Underwing coverts are mostly white, with dull brown spotting; axillars are mostly white. The iris is brown, the bill muted slaty to brownish-black; legs and feet are also slaty black to dark brown outwardly, but slate gray inwardly, and the webs are black. Females of *ferruginea* are generally darker than those of *jamaicensis*, the buff tones being largely replaced by more grayish tones. *Andina* females are intermediate in darkness.

Adult Nonbreeding Plumage, Females of jamaicensis

The female's nonbreeding plumage differs from the breeding plumage in that the crown is browner, the cheeks are off-white to pale brown, tinged with chestnut, and the facial stripe is less distinct. The innermost secondaries may be somewhat shorter and more rounded in profile. Body coloring otherwise is generally much like breeding plumage, but the upperparts are more speckled and sometimes barred with buffs and grays. Softpart colors are as in breeding plumage. Corresponding plumages of *andina* and *ferruginea* are still undescribed in detail, but at least those of the latter are not noticeably different from the breeding plumage.

IDENTIFICATION (FIG. 32A)

In the Hand.

Ruddy ducks are generally intermediate in bill length and bill width relative to the other *Oxyura*, although South American populations of this species approach the larger species in this regard. However, unlike the bills of two other large *Oxyura* (maccoa and white-headed ducks), the bills of ruddy ducks are not noticeably swollen basally, although the South American races approach this condition. Ruddy ducks from South America also have average maximum bill widths (usually 25 to 27 mm) and bill lengths (usually 40 to 45 mm) that are markedly greater than those of either the Australian or Argentine blue-billed ducks, both of which have very similar bill shapes and male plumages. Adult male North American ruddy ducks are easily distinguished in any season from all other *Oxyura* species by their white cheeks and contrasting dark "caps" that extend down to just below the eyes. Adult female North American ruddy ducks most closely resemble female maccoa ducks but are slightly smaller in wing measurements and slightly larger in average tail and culmen measurements. The less swollen bill and narrower cheek streaking of the North American form helps to separate females of these two otherwise very similar species.

In the Field.

In North America north of Mexico, field identification is simple; no other duck has such a long and usually cocked tail (especially in males) and such a short, thick neck. In the West Indies, Mexico, and Central and South America there are possibilities of local sympatry, and consequent confusion in the field, with masked ducks and Argentine blue-billed ducks. Masked ducks are appreciably smaller than ruddy ducks, have shorter and less distinctly "dished" bill profiles, and sometimes their white wing patches are partly exposed, even when swimming. They are also more prone to occupy heavily overgrown tropical swamps than are ruddy ducks. In the southern Andes and adjacent lowlands of Chile and Argentina, the Argentine blue-billed duck may provide a source of confusion as well. The Peruvian ruddy duck is substantially larger than the Argentine blue-billed duck and is much more likely to occur at higher elevations. Its bill also has a noticeably more swollen base, especially in males. Males of the Peruvian ruddy duck are more prone to swim with their tails partly cocked, even when they are not actively displaying, and females (plus nonbreeding or immature males) have much more obscured facial patterns than do Argentine blue-billed ducks. The male's feathered "horns" are not clearly evident, except during display, and are more poorly developed in the South American races than in the North American race. North American ruddy ducks are now locally resident in Britain as a result of escapes from captivity, and in continental Europe through recent dispersals. There they might possibly be confused with the native but considerably larger and generally paler white-headed duck, especially in southern Spain. The possibility of local hybridization between these two species is considered a potentially serious conservation problem for the threatened white-headed duck in Spain.

ECOLOGY

Habitats and Densities

Favored breeding habitats of North American ruddy ducks consist of permanent freshwater or alkaline marshes having extensive areas of emergent vegetation, open water, and a relatively stable water level. Suitable nesting habitats must have open water in fairly close proximity to nesting cover, including emergent plants that provide accessibility as well as adequate cover density and that can be bent down by the birds to form a nest platform. Water passages such as muskrat channels are needed to permit easy movement from the nest to open water when the nest is placed well back in a reedbed (Johnsgard 1975). Williams and Marshall (1938) reported finding 50 nests in a protected marshy area of 1200 hectares in Utah, or an average density of a nest per 24 hectares, with maximum densities of about a nest per 0.6 hectare in hardstem-bulrush *(Scirpus acutus)* habitat. Low (1941) estimated that in an Iowa breeding area of more than 400 hectares the average nesting density was about one nest per 8.5 hectares of marsh, and the maximum observed density was one nest per hectare, on a marsh of about 13 hectares. Bellrose (1980) judged that large areas of mixed-prairie habitats of the Dakotas and the aspen parklands and delta areas of western Canada may support an average density of about one breeding bird per 1.4 to 1.5 square kilometers (including uplands), with the densest breeding populations (averaging a bird per 0.7 square kilometer) occurring in the mixed prairie-parkland area of southwestern Manitoba.

Wintering habitats of ruddy ducks in North America are primarily brackish to slightly brackish estuaries or coastal lagoons of shallow depths. An abundance of submerged plants, small mollusks, and crustaceans are probably important as a winter food source (Johnsgard 1975). However, in Great Britain the birds almost exclusively winter on freshwater sites (Hughes, pers. com.). The majority of the North American population now winters along the Pacific Coast. This population consists of about 55 percent of all the ruddies seen during the midwinter waterfowl surveys made by the U.S. Fish & Wildlife Service during the period from 1955 to 1992 (a proportion that gradually increased from 48 percent during 1955–65 to 60 percent during 1982–92). This Pacific flyway population is mainly centered from California to western Mexico. California alone probably holds about 85 percent of the estimated 100,000 or more ruddy ducks that traditionally overwinter in the Pacific Coast states. These mainly occur in the San Francisco Bay area, but there are secondary populations in the Imperial Valley, San Joaquin Valley, and the southern coastal bays. Second to California as a wintering area is the western coast of Mexico, the birds mainly concentrating in coastal lagoons south of Mazatlan and near Acapulco. Very few ruddy ducks reach Guatemala in winter (Bellrose 1980).

On the Atlantic Coast, the Chesapeake Bay area has traditionally provided the best ruddy duck wintering habitats of the Atlantic flyway, typically with about 40,000 birds wintering there, but Pamlico Sound is also of considerable importance. Relatively few birds winter along the Georgia coast and Atlantic coast of Florida; in

Florida the ruddy duck is most common on inland waters and along the panhandle. Likewise, coastal Alabama and Mississippi support rather few birds in winter, although the interiors of these state receive higher usage. The Gulf Coast shoreline and nearby inland lakes of Louisiana and Texas (parts of the Mississippi flyway and Central flyway, respectively) collectively host average winter populations of about 20,000 birds. Coastal Texas often draws very large numbers of birds in early winter, a substantial number of which continue south to winter along the northeastern coast of Mexico, mostly north of Tampico (Bellrose 1980; Clapp, Morgan-Jacob, and Banks, 1982).

More recent continental winter surveys (see Reproductive Success and Status section below) have not differed greatly from those made during the 1950s and 1960s. For example, the collective winter North American ruddy duck population estimate has dropped from about 225,000 birds during the late 1950s to about 180,000 by 1992, representing an average population reduction trend of less than 1 percent annually.

Foods and Foraging

Several rather comprehensive studies on the foods of North American ruddy ducks now exist, most of which have been summarized earlier in various references (Johnsgard 1975; Palmer 1976; Bellrose 1980). In the most comprehensive study, that of Cottam (1939), which was based on the stomach contents of 163 adults, the majority (72.4 percent) of the food remains were of plant origin, with seeds and vegetative parts of pondweed, bulrushes, and wigeongrass *(Ruppia),* plus some algae *(Chara),* of greatest significance. The bulk of the animal remains were composed of midge (Chironomidae) larvae, with the larvae of horseflies (Tabanidae) of secondary importance (Table 8). In this study, as in all others involving only gizzard-content analyses, the proportion of soft animal materials such as insect larvae relative to hard plant materials such as seeds is seriously underestimated because of the rapid destruction of soft materials by the time they reach the gizzard. Using esophageal contents only, Siegfried (1973c) judged that 90 to 95 percent of the summer foods of adult ruddy ducks in Manitoba was composed of invertebrates, primarily dipteran larvae. Larval or pupal dipterans are evidently an especially important food source for juvenile birds, and animal materials in general are more important for younger birds than for adults.

During summer months, North American ruddy ducks feed almost exclusively on midge larvae or pupae (Siegfried 1973c; Gray 1980; Tome 1981; Woodin and Swanson 1989), and this same food source is also of primary significance during the winter (Hoppe, Smith, and Webster 1986; Euliss 1989). On some estuarine or coastal wintering areas such as Chesapeake Bay, snails and clams may tend to replace midge larvae as primary animal foods (Stewart 1962), although midge larvae are also consumed there. On some estuarine wintering sites benthic oligochate worms may also be utilized to a considerable degree (Stark 1978).

During late winter, aquatic invertebrates are the major foods of ruddy ducks using drainwater evaporation ponds of interior central California. They may compose more than 85 percent of the estimated total diet, judging from the esophageal contents of 27 birds. Of these, midge larvae and larval brine flies (Ephyridae, Diptera) were the most important component among all age classes and for both sexes (Hohman, Ankney, and Roster 1992).

Virtually nothing is known of the foods of the South American races, although Phillips (1922-26) reported that the stomachs of two Peruvian ruddy ducks from Lake Junin contained small bivalve mollusks (50 percent), amphipods (47 percent), a single corixid (Hemiptera) insect, and very small quantities of plant debris and seeds of various aquatic plants.

Foraging dive durations of adult and young North American ruddy ducks are similar to those of other *Oxyura* stifftails, as described earlier (Tables 6 and 7). Similar results for wild ruddy ducks were obtained by Siegfried (1973c, 1976b) in Canada and by Heintzelman and Newberry (1964) in New Jersey. When diving for food, ruddies propel themselves with simultaneous footstrokes until they reach the bottom; upon reaching the bottom, they retain their position just above it with similar simultaneous strokes directed perpendicular to the water surface. A fair amount of horizontal distance may be traversed between the point of diving and that of emergence (Siegfried 1973c). The return to the surface is done with little or no effort, the legs trailing behind as the bird's buoyancy quickly takes it back upward. While foraging at the bottom, the birds insert their bills into the silty substrate and rapidly open and close the bill while moving the head in a short lateral arc. The mandibular movement causes the soft substrate to be drawn into the mouth and forced out the sides of the bill, with the lamellae straining out edible materials. Although individual prey items are apparently thus not visually selected, the birds will swim toward aggregations of amphipods visible on the substrate surface and strain them from the water in the same manner as when feeding on benthos organisms (Tome and Wrubskiel 1988). Because of the nonvisually dependent foraging strategy of *Oxyura* stifftails, a considerable amount of feeding activity by North American ruddy ducks may actually occur at night (Stark 1978; Bergan, Smith, and Mayer 1989; Tome 1981) as is probably the case with maccoa ducks (Siegfried, Burger, and van der Merwe 1976).

Competitors, Predators, and Symbionts

Few direct competitors exist, at least in North America. Siegfried (1976c) analyzed foods and foraging behavior of ruddy ducks, canvasbacks (*Aythya valisineria*), redheads (*A. americana*) and lesser scaup *(A. affinis)* during summer in Manitoba. Ruddy ducks and canvasbacks tended to remain submerged for longer periods during foraging than did the other two species, but this difference is not necessarily indicative of competition effects. Although the four species had overlapping foods, each species tended to forage in a particular part of shared ponds and was more often found alone than with another species. Such usage of different foraging sites may be important in ecologically segregating those species whose diets overlap the most. The degree of dietary overlap of the ruddy duck with the other three species was judged greatest for lesser scaup, so selection for choosing different foraging sites may have been more important in reducing present-day interspecific competition than has been selection for choosing different foods.

The muskrat *(Ondatra zibethica)* is probably an important symbiont species for ruddy ducks in some areas, through the channels that it keeps open in reedbeds, allowing the ruddy duck ready access to these beds for nesting. Rarely, the ducks may even nest on old muskrat houses.

Probably the masked duck is only a minor competitor where it occurs, but in some parts of southern South America the Argentine blue-billed duck is certainly also present sympatrically (Hellmayr and Conover 1948; Johnson 1965; Casos 1992). In both cases the ruddy duck is substantially the larger of the species and certainly would be socially dominant. In the same general region (mainly Argentina) Peruvian ruddy ducks might be locally impacted by the presence of parasitic black-headed ducks. However, no clear indication yet exists that such parasitism of Peruvian ruddy ducks actually occurs, and black-headed ducks occur in generally warmer climates and at lower altitudes than do Peruvian ruddies.

Predators of both adults and young in North America are known to include minks *(Mustela vison)*. Ducklings are sometimes eaten by herons, including especially black-crowned night herons *(Nycticorax nycticorax)*, plus various larger gulls and some predatory fish such as pike and bass. Egg predators in North America certainly include larger gulls and various corvids such as ravens, crows *(Corvus spp.)* and black-billed magpies *(Pica pica)*. Egg or duckling predators probably also include mammals such as foxes *(Vulpes* spp.), skunks *(Mephitis* spp.) and raccoons *(Procyon lotor)*. In South America, caracaras *(Polyborus* and *Milvago* spp.) appear to be important egg predators for marsh-nesting birds. However, surprisingly few specific and important predators have been identified for ruddy ducks, perhaps because of the bird's remarkable ability to escape danger by diving.

ANNUAL CYCLE

Movements and Migrations

Long-distance migrations certainly occur in North American ruddy ducks, although they still remain fairly poorly understood.

Bellrose (1980) noted that migratory populations fluctuate wildly from year to year, with as many as 30,000 birds being seen in the Illinois River Valley in some years, and as few as 300 in others. Evidently, changing water conditions on the prairie breeding grounds may have a major effect on migratory routes and relative usage. Most migration probably occurs at night, although some very large diurnal movements have also been detected, involving combined migrations with large flocks of many duck species. However, most migratory flocks composed solely of ruddy ducks are probably quite small.

Bellrose (1980) illustrated several major migratory corridors in North America; the largest (involving an estimated 30,000 birds) were (1) from the northern plains states (North Dakota and western Minnesota) to the Chesapeake Bay area, (2) southward from the western Canadian plains to central Utah, and (3) thence southwestward from the latter area to the California coast. Fairly substantial corridors also extend (1) down the Mississippi Valley to western Florida and Louisiana, (2) through the central plains of the United States south to the Texas coast and east-coastal Mexico, and (3) from Utah and coastal California south to the western coast of Mexico.

Ruddy ducks begin leaving the northern prairie breeding areas in mid-September; they are nearly all gone by the end of October. Wintering areas in eastern North America acquire increasing numbers until early December; in the western states the numbers continue to increase through that month. The birds begin a slow movement out of wintering areas in February and continue to migrate almost through April. Ruddy ducks are among the last ducks to arrive on the prairie pothole breeding grounds of south-central Canada, the first birds typically arriving in southern Manitoba between late April and early May. The peak arrival occurs about the third week of May (Bellrose 1980).

Molts and Plumages

Some information on molts was provided in the Description section above. Hughes (1990) additionally reported that the molt of his captive birds into the first-winter plumage did not involve the wing feathers and that young North America ruddy ducks cannot be accurately aged until they first molt these feathers, which occurred in his study during the April following hatching, or when they were approaching a year old. Most juvenal wing feathers, especially the greater coverts, axillars, and inner secondaries, are relatively narrow and have tapered points, whereas the newly growing feathers in these areas are broader and blunter. The distinctively narrow and notched-tip juvenal rectrices also provide a good indicator of age until these feathers are molted, which in Hughes's captives occurred between August and November of their first year.

North American ruddy ducks studied by Carbonell at the Wildfowl Trust molted their tails twice annually; they also underwent two flightless periods annually. The prebreeding molt began during January, with the males first acquiring blue bills and then gradually attaining a chestnut body plumage. The remiges were then shed, and while flightless, the birds gradually lost their rectrices. Some individuals became tailless temporarily, although this was not invariable. In most cases the new rectrices grew between the old ones, in no clearly apparent sequence. The postbreeding molt began during August, with the birds losing their flight feathers first, and while flightless dropping their rectrices. Finally, the bright blue color of the bill was lost, and their body plumage became gray. Dominant males came into breeding condition before nondominant individuals did, and they were also the last to lose their breeding plumages. The remiges took about two weeks to regrow, the rectrices two to three weeks.

Breeding Cycle

At the Wildfowl Trust the first bubbling displays were seen by Carbonell (1983) as early as the beginning of January, and by early March all the males were actively displaying. Active courtship occurred until the end of July, and the dominant males continued to be sexually active until almost the end of August. The egg-laying period extended somewhat over three months, from late April until early August. In Utah, Joyner (1975) similarly found that the breeding season of wild North American ruddy ducks there lasted slightly over three months, with the first eggs being found in late April and the last being laid near the end of July. Low (1941) found nests as soon as early May in Iowa, as did Joyner (1975) in Utah, or about a month and a half after initial arrival. However, Bennett (1938) and Misterek (1974) observed egg laying in Iowa to begin in late May, which corresponds to other observations in Montana (Phillips 1922-26) and Manitoba (Hochbaum 1944). Additional comments on the breeding season of North American race may be found below (under Egg Laying and Incubation).

Relatively little is known of ruddy duck breeding cycles in the West Indies or in Central and South America. In Jamaica, nesting reportedly occurs from March to October (Downer and Sutton 1990), but in the more southern West Indies (Puerto Rico, Virgin Islands, etc.) it is said to extend from October to June, with the peak varying from year to year within this very broad time span (Raffaele 1983). Nesting by Colombian ruddy ducks may occur throughout most of the year, and downy young have been seen in December, March, May, and September (Delacour 1959). The breeding cycle of the Peruvian ruddy duck is evidently quite prolonged, varying with locality and altitude. Generally in the southern parts of the range (Chile and probably Argentina) it begins during September at lower elevations, during November in the mountains, and continues until January (Johnson 1965). However, in more equatorial regions such as at Lake Junin in Peru, breeding evidently is quite irregular. Some nesting there probably occurs throughout the entire year (Phillips 1922–26) but perhaps is concentrated between October and February (Fjeldså and Krabbe 1990).

SOCIAL AND SEXUAL BEHAVIOR

Mating System

Most evidence for wild North American ruddy ducks indicates that they temporarily form loose, sometimes nonmonogamous, pair bonds during the breeding season (Joyner 1975; Siegfried 1976c; Gray, 1980). At the Wildfowl Trust a short-term pair bond was apparently formed about one week before the first egg was laid by the female, but it lasted only until clutch completion and the start of incubation. Palmer (1976) described the ruddy duck's mating system as one of seasonal monogamy, but this is a misleading description of the actual bonding, which at most is brief and sometimes is polygynous. Gray (1980) estimated that most male ruddy ducks in California formed no pair bonds during the breeding season, but about a quarter of them attended a particular female for at least two days. Additionally, three individually marked males formed simultaneous pair bonds with two females. Further, two females formed sequential pair bonds with different males during the same breeding season, in one case prior to nesting and in the other after hatching a brood. Carbonell (1983) observed that in one pen at the Wildfowl Trust one dominant male was simultaneously paired to at least two, and sometimes three, of the females in that pen. Males at the Wildfowl Trust would seemingly be paired to more than one female simultaneously and would display to and copulate with other females as long as they were sexually active, as was also reported for wild birds by Siegfried (1976c). Ladhams (1977) also observed apparent nonmonogamous associations among feral North American ruddy ducks in southern England.

Territoriality

Male North American ruddy ducks do not defend a definable territory, nor confine their activities to areas having fixed boundaries. Rather, they tend to move about in search of females and follow them in the selection of nesting sites (Siegfried 1976c). However, males do defend an indefinite area of about 3 meters surrounding their female mates, which Joyner (1975) has described as a "revolving territory." Joyner (1969) earlier suggested that males may defend a small area several meters in diameter surrounding the nest site and judged that such apparent nesting territories may be spaced from about 6 to 30 meters apart, the distance depending on the presence or absence of channels through the emergent vegetation. The more this vegetation is disrupted by channels, the closer the apparent territories might be situated. Alternatively, such channels may simply allow greater access to females for locating nest sites, and thus not be directly related to territorial tendencies at all.

Courtship and Pair-Bonding Behavior

Several accounts of the displays of the North American race of this species have been published, of which the ones by Johnsgard (1965a) and Palmer (1976) are among the more complete. Behavior by feral birds in England was described by Ladhams (1977). Early work on the species was done by Helen Hays, who first named most of the male displays. Unfortunately, her work was never published, though it has been well summarized by Palmer. Carbonell's (1983) observations were based on up to five males of the North American race at the Wildfowl Trust. Johnsgard has made observations on large numbers of wild and captive birds of the North American race, several captive males of the Peruvian race, and a few first-generation hybrids between the Peruvian race and the North American race. Displays of the Colombian race are as yet undescribed in detail, and unless otherwise indicated the following account relates specifically to the North American race.

The average duration of display bouts of the North American ruddy duck at the Wildfowl Trust was 3 minutes, and the maximum amount of time spent displaying by a male without interruption was 14 minutes. During the breeding season males at the Trust spent about 7 percent of their daylight activities in sexual display (Table 3) (Carbonell 1983). Hughes (1992) estimmated that wild males in England spent an average of about 4 percent of their diurnal time in courtship display during the courtship period; his study period totaled more than 107 hours of observations.

In both the Peruvian and North American races, males who have chased another male will frequently return to the female in a *rush* posture. In this posture the male swims very fast, with his breast above the water, his bill against the breast, and the back of his body almost submerged in the water. This or a similar posture occurs during the final phase of the ringing rush. Contrary to Palmer (1976) and Ladhams (1977), this sexually oriented rush display is not at all comparable to the aggressive hunched rush, and the scapulars are not raised during the rush display.

In the aggressive *hunched rush,* which is a more intensive version of the "hunched threat" (Gray) or "threat swimming" (Hays), the scapulars are strongly raised as the male aggressively approaches another individual (Fig. 19F). A counterpart of the male's hunched-threat display is also performed by females as part of their agonistic responses to courting males, in which the female swims with her scapulars raised, her head lowered, and her bill open, an "open-bill threat" posture (Fig. 19G), a posture that has also been interpreted as representing inciting behavior. The highest level of aggressive

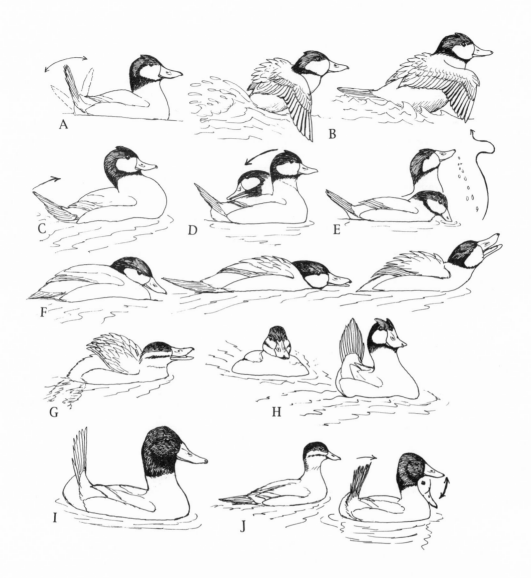

Figure 19. Social-sexual behavior of the male North American ruddy duck:
(A) tail cocking, (B) ringing rush, (C) tail cocking following ringing rush, (D) cheek
rubbing, (E) bill-dip, head-flick, and (F) threat swimming. (G) Female North
American ruddy in threat posture and (H) male tail flashing while leading female.
Also shown is the male Andean ruddy duck's (I) tail-cocking and (J) bubbling display. (After sketches from cine sequences by Carbonell, photos by Johnsgard, and
sketches in Palmer 1976.)

male-male interactions are bill-to-bill faceoffs in the hunched-threat posture, which are usually followed by a diving retreat on the part of one or by actual vicious biting, wing-beating, and clawing attacks as the birds stand face-to-face almost upright in the water amid a spray of water. Hughes (1992) found that paired males were involved in significantly longer aggressive interactions than were single males and were more prone to perform hunched rushes and hunched threats than were single males, but were less likely to perform open-billed threat displays.

The *ringing* (or *ring*) *rush* ("display flight" of Miller, McLandress, and Gray 1977 and Gray 1980) is an important but infrequent display. It may occur independently (Fig. 19B) but also often directly follows a bubbling sequence (Fig. 20). It consists of a half-swimming, half-flying rush across the water (Fig. 20, sketches K–Q), which often places the male closer to a female or a courting group. The ringing rush is so-named because of the characteristic ringing sound that is generated by the feet and perhaps also by the wings striking the water. It typically ends with a skidding stop near the female (Fig. 20, sketch Q) and is quickly followed by a return to the tail-cocked posture (Fig. 19C). From 5 to 15 wingbeats typically occur during the rushing phase (Ladhams 1977), although the bird never actually rises above the water surface. A similar, nondisplay behavior involving running across the water while flapping the wings sometimes occurs as a result of limited disturbance; this kind of escape behavior was called "skeetering" by Hughes (1992).

Ringing rushes are most commonly performed under conditions of intense male sexual competition for females. Gray (1980) noted that these "display flights" are usually directed toward a female, regardless of whether the male is still unpaired (76 to 78 percent of the observed displays were so oriented) or is already paired (83 to 93 percent). However, at times these displays are performed by single males without any apparent specific stimulus; they are also sometimes directed away from opponents, following hostile interactions. However, Gray noted that most of the ringing rushes performed by paired males were directed toward their mates and often occurred after aggressive interactions with other males. Unpaired males typically performed ringing rushes when approaching females, with or without accompanying bubbling displays. Gray noted that ringing rushes were the least frequent of the displays that she quantified; she nevertheless observed sequences consisting of up to as many as 25 separate rushes. She observed a maximum display rate of two rushing sequences in 10 minutes, but often there were fewer than two such sequences per hour.

The *bubbling* (or "bubble") sequence (Fig. 19J; also sketches A–J in Fig. 20) is the major sexual display of ruddy ducks and composed 81 percent of the total number of displays tallied by Carbonell. The duration of 40 sequences of the North American race was 0.7 to 1.7

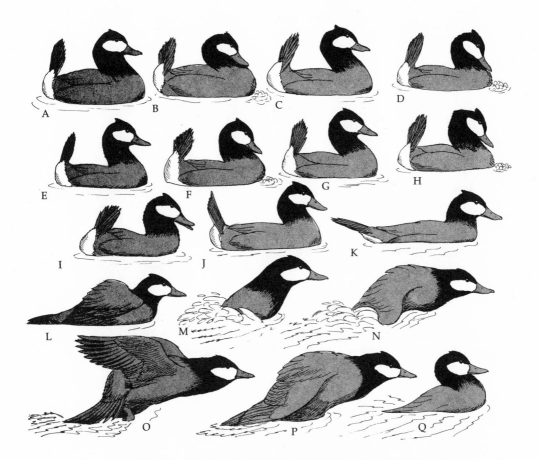

Figure 20. Display sequence of bubbling (A–I) followed by ringing rush (J–Q) in the male North American ruddy duck. (After a cine sequence by Johnsgard.)

seconds, averaging 1.26 seconds. Besides being a primary courtship display, bubbling also sometimes occurs when the birds are suddenly and slightly alarmed. In that same situation it also occurs in females and additionally has been observed in unfledged ducklings. The bird begins by raising his head and tail to a vertical position (also called "tail-cocking" or "head-high" posture) (Fig. 19A, I). In this "ready" position the feathered "horns" are erected (these are not apparent in the Peruvian race), and the tail is typically closed but may sometimes be spread slightly. The neck is distinctly enlarged (visibly more so in the North American race than the Peruvian form). Then the bird begins to beat the underside of his lower mandible against the lower part of the inflated neck, producing dull tapping sounds and generating a ring of bubbles from the base of the breast, as air is forced out from between these feathers (sketches B-H, Fig. 20. As the sequence proceeds, the tail is brought increasingly forward, so that at the end of the bubbling sequence, as the bill is opened and a weak, belching call uttered, the tail is held at an angle diagonal to the back (sketch J, Fig. 20). The scapulars are also momentarily raised and the neck is brought forward briefly at the end of the bubbling sequence.

The number of billbeats observed by Carbonell during a bubbling sequence in the North American race varied from 5 to 9, with an average of 6.9 and a mode of 7 in a sample of 40. In a first-year male the observed range was 5 to 9, with an average of 6.7 and a mode of 6 in a sample of 10. In this case the mean duration was about the same (1.26 seconds), but the actions were not so clearly stereotyped. In a sample of 15 bubbling sequences by the Peruvian ruddy duck, the observed range of bill beats was 2 to 4, with an average of 3.0, and in 18 bubbling sequences by first-generation hybrids of the Peruvian and North American races the range was 5 to 8, with an average of 7.0 (Johnsgard pers. obs.). Ladhams (1977) stated that more than 40 bubbling displays may be performed by a single male within a 20-minute period; thus at least two bubbling displays per minute are frequently performed. Gray (1980) observed that bubbling displays are performed at about the same rate throughout the daylight hours and that the displays extended from prenesting to hatching periods without much diminution, with single males averaging 11 display bouts (series of displays averaging less than 30 seconds apart) per hour in the prenesting period of 1975, versus 6 per hour during the hatching period. Corresponding figures for paired males during these two periods of 1975 were 6 and 4 display bouts per hour.

Gray (1980) observed that bubbling is primarily evoked in social rather than nonsocial situations (such as male advertisement), that unpaired males displayed most frequently prior to the nesting period, and that they oriented most (71 percent) of their bubbling displays toward females. Carbonell observed the bubbling display to be directed only toward females, except for one case of a homosexual pairing. Johnsgard has seen it oriented toward males on occasion and believes that it serves a hostile function in such cases. When bubbling is stimulated as a result of alarm, no clear orientation is evident. Gray came to similar conclusions with her observations of wild birds and judged that the vigor and persistence of the bubbling display might reflect dominance relationships among individuals. After pair bonds have been formed, the rate of bubbling diminishes slightly, and in this postpairing situation the display may serve to help maintain pair bonds. Single males displayed more in nonhostile situations than did paired ones, and paired males directed most of the bubbling in hostile situations toward their mates. Most observed bouts (71 percent) of bubbling were directed toward the female, regardless of the male's paired or unpaired status, and in the case of unpaired males a significant proportion (24 percent) were performed as the male followed the female. However, among paired males, 18 percent were performed as the female followed the male (Gray 1980). This coordinated activity, with the female following the male as he swims ahead with his tail cocked ("tail flashing" of Hays and Gray) and his nape directed toward her (Fig. 19H) is seemingly a good indication of completed pair bonding in ruddy ducks and somewhat resembles the combination of female-inciting, male-leading behavior that is an important pair-bonding mechanism in waterfowl generally (Johnsgard 1965a). Hughes (1992) included tail flashing within his "courtship alert" posture, in which the orientation of the tail coverts varies with respect to the female.

The *bill-dip, head-flick* display (Fig. 19E) represented 5.6 percent of the total activities tallied, and individual components lasted 0.4 to 0.5 second (average of five observed, 0.42 second). In this action, the male, with his "horns" raised and tail held at a diagonal angle, dips his bill in the water and then quickly withdraws it and shakes it laterally while lifting it above the horizontal. The display is closely similar to the corresponding comfort movements. The display was only observed in paired males and is probably essentially an invitation to copulation, although females sometimes seem to ignore it.

Head dipping is a brief (but unmeasured) and often repeated action that represented 4.4 percent of the total activities tallied during social display. It is very similar to the corresponding bathing movement. It was only observed in paired males and is of uncertain social significance.

Wing shaking is another brief (but unmeasured) action and represented 4.0 percent of the total activities tallied by Carbonell during social display; it is very similar to the corresponding wing-shaking comfort movement. Wing shaking was observed by Carbonell only among paired males and was once seen during a precopulatory sequence of bill dipping and head flicking. Wing shaking probably corresponds to the "wing buzzing" described by Hughes (1992). It consists of rapid "buzzing" wing movements and apparently represents a ritualized version of wing shivering as described by Siegfried (1973a). The "wing quivering" described by Ladhams (1977) probably also represents comfort-movement behavior.

Head shaking, yet another brief (but unmeasured) action, represented 3.3 percent of the total activities tallied by Carbonell during social display and is very similar to, if not identical with, the corresponding comfort movement. It was almost always performed after a bubbling sequence. Head shaking was considered by Palmer (1976) to be a flight-intention movement and has been observed in this context by Hughes (pers. com.).

Cheek rolling (or "rolling cheeks on back") is a brief (but unmeasured) action that represented 1.7 percent of the activities tallied by Carbonell during social display; it is very similar to, if not identical with, the corresponding comfort movement. In spite of the low incidence of occurrence it did seem to be ritualized activity. It was performed only by paired males, and only when no other males were present. Cheek rolling was also described by Palmer (1976) as a flight-intention movement.

Observed by Palmer (1976), pseudosleeping actually consists of nonritualized resting behavior, which is typical of and commonly performed by both sexes of stiff-tailed ducks in resting situations. (See Fig. 26H.) Contrary to some accounts, Johnsgard and Carbonell both believe that it should not be regarded as an appeasement display in ruddy ducks. Similarly, the corresponding "bill-hiding" or "turning-the-back-of-the-head" posture illustrated in Palmer (1976) is only very questionably an appeasement gesture in this or any other *Oxyura* species.

Fertilization Behavior

Six complete copulation sequences were observed by Carbonell during her studies at the Wildfowl Trust, and numerous copulations have been seen in captive and wild birds by Johnsgard. In one sequence seen by Carbonell, the female was diving while the male was performing bubbling displays. After one such dive, she sank her body in the water, whereupon the male stopped displaying and immediately mounted her. On a second occasion the female approached a resting male, performed a head shake, and then became prone. The male became alert, assumed a head-high posture, and mounted her. On the other four occasions the male performed repeated head-dipping, head-flicking ("bill-flicking") displays while the female was diving. Eventually the female stopped diving, approached the male, and became prone. Copulation lasted 3 to 5 seconds; after treading, the male performed up to three bubbling displays before starting to preen. These latter sequences correspond fairly well to the typical copulation sequences as described by Johnsgard (1965a) and to those observed in wild birds by Joyner (1969). They also correspond to four sequences observed by Hughes (pers. com.), although Hughes noted that the female did not assume a prone position before copulation began.

Rape behavior among North American ruddy ducks is relatively uncommon at the Wildfowl Trust, being noted only twice by Carbonell. However, Gray (1980) observed numerous cases among wild birds in California. She noted that it was usually preceded by a "flight chase," in which the male ran along the water surface while pursing the female. Splashing associated with this chase sometimes attracted other males, who then joined in the pursuit. Most females (22 of 25) subjected to rape attacks attempted to escape by diving and swimming away under water, but in three cases they attempted to defend themselves. In two of these cases the females were already paired, and their males helped to defend them. All of the observed rape attempts occurred during the nesting period, when most females were either laying eggs or already incubating. Most (84 percent) of the 25 females involved in attempted rapes were unpaired, and 25 percent of such rape attempts were successful. One of four attempted rapes involving paired females was also successful. Both paired and unpaired males were observed to participate in rape attempts.

NESTING AND PARENTAL BEHAVIOR

Nests and Nest Building

Of 42 North American ruddy duck nests found by Carbonell at the Wildfowl Trust, 57.1 percent were in nest boxes, with the rest placed among rushes (14.3 percent) or among other types of herbaceous vegetation (28.6 percent). Nests built among vegetation were partially covered by a roof made with bent-over stems. All of the nests were within a meter of water and lacked down. Measurements of these nests are shown in Table 12.

Among the numerous studies of wild North American ruddy ducks, Williams and Marshall (1938) found that 32 percent of 50 nests were in hardstem bulrush, representing a proportionate plant usage that was greater than that species' relative abundance. About the same number (30 percent) were in the much more common alkali bulrush (*Scirpus paludosis*). Twenty percent were in saltgrass (*Distichlis*), which is not an emergent species but rather is a shoreline grass, and 14 percent were in cattails.

Of 30 nests found by Misterek (1974) on Rush Lake in Wisconsin, 70 percent were located on floating cattail islands in midlake. Only 16.6 percent of the nests were on shoreline cattail mats, and 13.3 percent were on midlake beds of hardstem bulrush. Eighteen of the the nests were in medium cover, and 12 in dense cover. Forty percent of the nests were no more than 9 meters from another ruddy duck nest, and the closest were located within 2 meters of each other. Interestingly, the nests located on the favored floating cattail islands also had the highest hatching success (62 percent), those on shoreline cattail mats had an intermediate rate of success (40 percent), and those in hardstem bulrush beds the lowest (25 percent). The nests in drier sites had shallow water-filled trails leading to open water, which was within about 2 meters on average. In an Iowa study (Bennett 1938), 14 of 22 nests were in stands of roundstem bulrush (*S. occidentalis*), a species apparently favored for nesting because of the relative ease with which it can be bent over to form a nest. Low (1941) concluded on the basis of 71 Iowa nests that nesting cover was not determined so much by preferences for specific plants as for plant cover associated with a suitable water depth at the time that nesting was begun. A water depth at the nest site of about 0.25 meter (10 to 12 inches) is evidently favored, and no nest was located in water a meter or more deep. In this study, lake sedge (*Carex lacustris*) was the most favored cover type (especially during a year of high water levels that flooded shoreline vegetation), followed by hardstem bulrush and cattails. On the basis of actual plant usage relative to total available plant cover, the nesting preferences in decreasing sequence were slender bulrush *(S. heterochaetus)*, whitetop *(Scolochloa)*, hardstem bulrush, lake sedge, and narrow-leaved cattail *(T. angustifolia)*. Some instances of dryland nesting by ruddy ducks have been reported (Keith 1961; McKnight 1974). Although only 1 of 15 nests studied by McKnight was over water, all nests were situated within 0.9 meter of water, the mean distance to water being only 0.4 meter. Featherstone (1975) has also documented the tendency for this species to nest preferentially in dense vegetation.

Egg Laying and Incubation

North American ruddy ducks at the Wildfowl Trust began laying eggs during the second half of April. The last eggs were laid during the first week of August in two years, and during the last week of June in the third. Palmer (1976) summarized North American nesting phenology information by stating that in the southern states clutches might be expected by late April or early May, whereas in areas farther north they might not be laid until late May or early June, with the last clutches occurring as late as early August. Bellrose (1980) estimated an overall egg-laying period of only 53 to 63 days as typical for most breeding areas of North American ruddy ducks, which would certainly not be long enough to allow for rearing second broods. Newly hatched broods have been seen in Texas as late as mid-September. This represents a potential maximum egg-laying season of about four months in the southern parts of the U.S. range, which would barely allow enough time for two broods to be hatched and raised successfully. So far, good evidence for double-brooding by any species of stifftail under natural conditions is apparently lacking.

During the period of follicular growth, female North American ruddy ducks at the Wildfowl Trust spent 47.5 percent of the day foraging, as compared with 27.1 percent during the egg-laying period. However, it is probable that females spent longer periods feeding during the day the first egg was laid, and on the day before. The highest average female body weights of the year occur in ducks just as egg laying begins (see Weights above), and it then quickly declines. As the laying period progressed, the females spent progressively less time feeding. While incubating, females spent an estimated 92.4 percent of the daytime period on the nest; Siegfried, Burger, and van der Merwe (1976) reported a somewhat lower percentage (76 to 83 percent) of nest attentiveness for wild birds, representing about 19 hours per 24-hour cycle, with similar attentiveness rates whether under dark daylight conditions. Similarly, Gray (1980) reported that incubating females spent an average of about 85 percent of the 24-hour diel cycle on the nest. Tome (1991) likewise estimated a 73.5 percent resting (mostly incubating) incidence during diurnal portions of the incubation period, with foraging activities occupying most of the female's nonresting activities and recess periods averaging about 40 minutes each.

Clutch-size data for North American ruddy ducks have been summarized in Tables 14 and 15, both for captives at the Wildfowl Trust and for wild birds as determined by Joyner (1975). Clearly, great variation exists in these data, the probable result of a combination of factors involving individual (or age-related) variation, dump-nesting, and parasitic egg laying. However, 7 to 8 eggs would seem to be typical for clutches presumably produced by individual wild females, as compared with 6 for captive females at the Wildfowl Trust. Thus Bennett (1938) determined a mean clutch of 7.05 for 18 wild nests, and Peck and James (1983) summarized data for 18 Ontario nests, which averaged 7.2 eggs. Low (1941) reported an average of 7.6 eggs for 71 wild nests, and Misterek (1974) an average of 8.4 eggs for 12 nests (excluding known parasitically laid eggs). Bellrose (1980) summarized data from various field studies to produce a collective mean clutch size of 8.05 eggs for 312 nests.

The incubation period for North American ruddy ducks at the Wildfowl Trust was determined by Carbonell to be 24 to 25 days, which is in the general range (23 to 26 days) established for wild birds (Low 1941; Joyner 1969; Palmer 1976). Losses during incubation are often caused by water fluctuations (Low 1941), which result either in nest destruction through flooding or in nest desertion as the nest becomes increasingly stranded on dry land. Predation by fairly aquatic mammals such as minks may be a cause of some nest losses. However, most wild nests are effectively isolated from land predators because of their aquatic locations and are sufficiently hidden overhead by vegetation to avoid detection by most avian predators. High levels of ruddy duck nesting success on mammal-free islands have been reported by Keith (1961) and on similarly isolated floating cattail islands by Misterek (1974). The average interval before a second clutch was laid by a female after her first clutch had been removed was 23 days at the Wildfowl Trust; the range for three such cases was 19 to 30 days (Carbonell 1983). The incidence of renesting in wild birds is still rather uncertain, but Siegfried (1976a) believed that it occurs only infrequently, as was also concluded by Bellrose (1980). One case of renesting was documented by Tome (1987).

Hatching and Brood-Related Behavior.

One female and her brood were followed by Carbonell (1983). This female had a clutch of six eggs, four of which hatched on one day, the fifth the following day, and the sixth a day later. The eggshells were not removed from the nest but, rather, were crushed by the adult. The female and her first four chicks returned to the nesting box the first night after hatching and again the second night, in contrast to Joyner's (1975) belief that the hens do not return to their nests after hatching.

It required about 15 minutes for the first four chicks to exit the nest, which began when the female came out of the nest box and began calling with a "gek-gek" sound. After all were in the water they began to feed immediately, straining food from the water surface. The female swam nearby, constantly watching them and uttering the same soft call. The chicks followed her closely, uttering soft "peep-peep" notes and feeding directly behind her. The first trip away from the nest lasted nearly two hours and occurred in late afternoon. During the first week the main activity of the chicks was feeding (44.6 percent of time); sleeping occupied 19.6 percent (Table 5). After the sixth day, the brood did not return to the nest box and slept elsewhere thereafter. The female's primary activity during the first week consisted of being alert and watching for disturbance or predators (35.3 percent) while swimming (Table 4). The ducklings did not begin diving for food until their second day. The length of the dives by ducklings during their first several weeks of life (Table 7) is somewhat longer than the average diving durations reported by Joyner (1975, 1977a) for wild ducklings in water of comparable depths.

When the female detected any disturbance she would call louder and faster to her brood and stretch her neck in a head-high posture. The chicks would then gather and stay close to the female until the cause of the alarm was over. With sudden alarm the chicks would dive quickly and then cluster after emerging, as was also observed in wild birds by Joyner (1975). If the alarm was sudden, the female would sometimes perform the head-high, tail-cock, or (once) the bubbling display, as has also been reported by Joyner (1975, 1977a) and Siegfried (1977). A rudimentary version of the bubbling display was once observed by Carbonell in an alarmed duckling, as has also been observed by Johnsgard (1965a). Gray (1980) observed that the sudden appearance of a predator such as a gull would cause the young to cluster around the female's erect tail and white undertail coverts and then swim to cover. Head bobbing by the female would also apparently attract the young to her, would stimulate diving, or would initiate a move to cover. At times small ducklings may also try to associate with males (Ladhams 1977).

Females threaten intruders in an open-bill threat posture (Fig. 19G), which sometimes becomes a chase in the hunched-rush posture. Gray (1980) once observed a "distraction" display when the ring-billed gull (Larus delawarensis) flew near a female with ducklings. After the young dived, the female began swimming in small circles, her body low on the water surface, flapping her wings and vocalizing. She then moved to cover, followed by the young, diving repeatedly. There are numerous accounts in the literature of males associating with females leading broods, and some of these have attributed brood defense to males (e.g., Misterek 1974). However, there is little evidence of any effective protective role for males in the experience of Carbonell, Johnsgard, Joyner (1975, 1977a) or Siegfried (1977); instead, the association seems to result from a prolonged sexual attraction of the male to the female. Gray (1980) observed 7 rape attempts during the hatching period, in addition to 18 observed earlier in the prenesting and nesting seasons.

Carbonell observed that as the chicks grow older the amount of time they spend feeding gradually decreases (41.1 percent the second week, 39.1 percent the third), and the amount of time spent resting correspondingly increases. Joyner's (1975) observations that younger ducklings feed more frequently but for shorter periods than older ones corresponds with those by Carbonell at the Wildfowl Trust. The close association between the female and the ducklings gradually fades, and at 24 days after hatching the female and brood studied by Carbonell moved to another pen, where the female began her postbreeding molt. By that time the young were almost entirely independent of their mother, as Joyner (1975, 1977a) similarly reported for ducklings four or five weeks of age. Similar accounts of increasing duckling independence with advancing age have been provided by Beard (1964), Misterek (1974), Siegfried (1977), Gray (1980), Palmer (1976) and Hughes (1992). The diverging needs of the mother (increased foraging activities to restore energy used in breeding and to prepare for postbreeding molting stresses) and the relatively high ability of the young to look after their own survival requirements no doubt facilitates this early brood dissolution process.

The fledging period for wild birds has been variously estimated as 6 weeks (Palmer 1976), 52 to 66 days (Hochbaum 1944), and 8 weeks (Joyner 1975). Wild ruddy ducklings in Manitoba were found to attain their adult weight in only 42 days (Bellrose 1980). At the Wildfowl Trust fledging occurred at approximately 8 to 9 weeks (Carbonell 1983).

REPRODUCTIVE SUCCESS AND STATUS

Some data on the hatching success of wild North American ruddy ducks nests have been summarized in Table 19. Nesting success rates (defined as the percentage of nests hatching at least one duckling successfully, relative to total nests under observation) range from 34 to 73 percent and average about 55 to 60 percent. Hatching success rates (defined as the percentage of eggs hatching successfully, relative to total eggs in sample) ranged from 31 to 73 percent, averaging about 50 percent. This would suggest that an average of about four ducklings are hatched per nesting female per breeding seson. Because of brood-size instability, little information can be obtained in the field on probable rates of prefledging duckling mortality, which normally represents the period of greatest posthatching waterfowl mortality. There is also no quantitative information on annual adult mortality rates. However, as mentioned in Chap. 5, the recruitment rate of juvenile birds into the fall ruddy duck population is probably about 50 percent, which is likely to approximate the collective annual mortality rate for the entire wild ruddy duck population (immatures plus adult age classes). In most species of North American ducks that are hunted for sport these annual mortality rates typically average about 60 to 70 percent for first-year immatures and 35 to 45 percent for adults (Bellrose 1980).

Population sizes of the two South American races are still undetermined. However, that of the endemic Colombian race in particular should be monitored if possible, as it is inherently small and might easily become extinct before it has even been studied in any detailed manner. Fjeldså (1986) suggested that this race might actually be a "relict" hybrid population, resulting perhaps from a Pleistocene contact between *jamaicensis* and *ferruginea*, with the majority of the males now tending toward one plumage extreme or the other. A recent inventory of wetlands in Colombia (Scott and Carbonell 1986) indicated the presence of this race on the following sites in the Departamento de Cundinamarca: Embalse de Neusa, Bañados de la Florida, Laguna de Pedropalo, and Laguna de San Ramon (18 adults plus young in 1981). Additionally, 15 to 25 pairs were present on the Laguna de Tota (Boyaca Department) and about 80 birds have recently been observed on Laguna del Otun (Risaralda Department). They have also been reported from Lago Yaguarcocha, Imbabura Province, Ecuador.

The now-protected Lake Junin population of the Peruvian ruddy duck is an important stronghold of that race; its population evidently consists of several thousand birds. Lake Titicaca holds unknown numbers, but it would seem to be a potentially important area as well, as it has good stands of *Typha* and other emergents (especially *Schoenoplectus*) as well as many edible species of floating and submerged vegetation. Other Peruvian sites where this race has recently been reported include Lagunas de Marcapomacocha (Pasco and Junin departments), Lagunas de Huacarpay, Pomacanchi, Asnacocha, Pampa Marca Cuzco Department), Sallahu and Lagunillas (Puno Department), and Yaurihuiri (Ayacucho Department). An estimated 2500 ruddies of this race occur on Laguna Alalay and about 500 on Laguna Corani, both in Departmento de Cochabamba, Bolivia. Others also occur in Bolivia on Lago Urur-Uru, Departmento de Oruro (Scott and Carbonell 1986). The birds seemingly are fairly common on the Andean lakes of central Chile and Argentina and on at least some of the adjoining temperate low-elevation marshes and lakes to the south.

Populations of the North American ruddy duck have been fairly closely inventoried for nearly four decades, mostly through winter surveys and breeding season counts. Through the assistance of

David E. Sharp of the U.S. Fish & Wildlife Service, these data on ruddy ducks have been assembled and made available to the authors. They include their estimated breeding populations (1955–92 spring waterfowl breeding population and habitat surveys), their estimated winter populations (1955–92 midwinter waterfowl surveys), and total hunter-harvest estimates for ruddy ducks in the United States (1961–91) and Canada (1971–91). The overall 38-year average breeding population of ruddy ducks was estimated at 382,908 birds, with a slight long-term upward population trend statistically indicated. The midwinter waterfowl survey had an overall 38-year average estimate of 193,859 birds, but with a comparable slight downward long-term trend apparent. Collectively over this entire period, the Pacific flyway has supported 54 percent of the total wintering population, the Atlantic flyway 31 percent, the Mississippi flyway 11 percent, and the Central flyway 4 percent.

The hunter-harvest estimates show a much clearer trend. After figures rose during the 1960s, a sharp downward trend has been apparent since 1971, the first year for which combined U.S. and Canadian figures are available. The long-term (1961-91) average U.S. estimated annual hunter kill was 52,889 ruddy ducks; the 1971–91 Canadian harvest estimate was 4127 birds annually. The highest combined U.S. and Canadian kill occurred in 1972, when an estimated 122,506 birds were shot by sport hunters. By comparison, the lowest kill occurred in 1990, when only 10,361 were estimated to be shot. These considerable changes in harvest may reflect hunter-value attributes—that is, hunters' perceptions of the relative desirability of ruddy ducks as trophys—as much as the average long-term availability of ruddy ducks to hunters.

Comparing the collective hunter-kill figures against the collective winter survey totals, it may be estimated that perhaps as much as 30 percent of the ruddy duck population may be harvested by hunters each year. However, if the breeding-grounds estimates are used, this percentage drops to about 15 percent. Neither figure would suggest that overhunting has been a significant factor in long-term ruddy duck population changes.

In short, two of these three sets of statistics suggest possible slight (midwinter surveys) to sharp (hunter-kill data) long-term declines in North America ruddy duck populations, whereas one set (breeding population surveys) suggests that the continental population may actually be increasing slightly. In any case, as of the early 1990s there would appear to be several hundred thousand birds in the North American population, making its survival apparently secure at the present time. Yet other relatively unhunted North American waterfowl populations, such as that of the spectacled eider *(Somateria fischeri)*, have plummeted in recent years with no warning and no clear reason. Thus the North American ruddy duck's status is by no means as secure as it might seem. Several major kills of ruddies have occurred in wintering areas as a result of oil spills (Johns-gard 1975), and the concentration of a good share of the continental wintering ruddy duck population in places like San Francisco Bay makes this a continuing threat to the species. Several cases of oil spills involving ruddy duck mortality were summaried by Clapp, Morgan-Jacob, and Banks (1982).

Audubon Christmas Bird Counts provide useful independent data on American winter bird distributions but have not been thoroughly analyzed since Root (1988) did so for the 1962–63 to 1971–72 decade period. She found peak usages by ruddy ducks to be centered in southern California (Imperial Valley, near the Salton Sea Refuge) and in an area of the lower Mississippi River along the Arkansas-Mississippi boundary, between the Yazoo and White River national wildlife refuges. In contrast to aerial surveys, Christmas counts tend to underestimate coastal use by the birds and overestimate inland use as a result of limited visual access to coastal waters by bird-watchers. Since 1972 the highest reported counts of ruddy ducks in these surveys have been in California on ten occasions, in Texas on three, in Pennsylvania twice, and on single occasions in Maryland and Florida. Specific areas reporting the highest single North American counts since the early 1970s include Palo Alto, California (five times), Salton Sea, California (four times), Laguna Atascosa refuge, Texas (three times), Glenolden, Pennsylvania (twice), Hayward, California (once), Kern River valley, California (once), Tampa, Florida (once), and Baltimore harbor, Maryland (once). These counts have averaged nearly 20,000 birds annually (maximum 52,100, Laguna Atascosa, 1975).

The North American ruddy duck is now well established as a feral species in Britain. After initially escaping from captivity at the Wildfowl Trust in 1952, the species has continued to increase in southern England, its population reaching 100 birds by 1974 (Ogilvie 1975a). During the late 1980s and early 1990s the British population was increasing at a rate of about 9.4 percent annually and was judged likely to continue to increase indefinitely because of many still-uncolonized sites in the country. Apparently, the species is having little direct deleterious effects on native waterbird species. Its United Kingdom population as of 1992 consisted of about 3500 birds, including about 600 breeding pairs (Hughes and Grussu, unpubl. ms.).

Since its range expansion to mainland Europe (first reported from Sweden in 1965), it has been seen in at least 15 countries as of 1992, with a maximum number of wintering birds (about 200) and breeding records (about 10 through 1992) reported for the Netherlands. The Netherlands population includes one to two breeding pairs and 25 to 30 birds wintering annually. Breeding has also been reported in Iceland, France, Belgium, and Spain and is suspected in Ireland. There have also been recent reports of the species' occurrence in Morocco and the Ukraine (Hughes & Grussu, unpubl.).

The ruddy duck reached Spain by 1982, and as of 1992 there were at least 81 records for that country. Additionally, there have been at least 20 confirmed ruddy duck and white-headed duck hybrids (including second-generation individuals) observed as of 1992. This hybridization, together with the effects of interspecific competition, is regarded as posing a serious threat to the endangered native Spanish population of white-headed ducks. Males of ruddy ducks and of interspecific hybrids are more aggressive than are male white-headed ducks, placing the latter species at a social disadvantage during courtship (de Vida 1993). The threat to the remnant European population of the globally threatened white-headed duck posed by competition and hybridization with the self-introduced ruddy duck was explored at an international workshop held in England during the spring of 1993 (Anon 1993).

Male maccoa duck performing vibrating trumpet toward female.

Maccoa Duck

OXYURA MACCOA
Smith 1837

Other Vernacular Names
> African ruddy duck, red diving duck; Maccoa-ente (German); erismature maccoa (French); Macow-eend, Makou-eend (Afrikaans Dutch, probably originally derived from a native word "kacauw" or "makou" for a kind of goose, presumably the spur-winged goose, which is sometimes called the Wilde-makou in Afrikaans); pato maccoa (Spanish).

Range of Species (Fig. 21)
> Widely scattered in two isolated populations of eastern and southern Africa. The more northerly population is local in Ethiopia and more common on the rift valley lakes of Kenya, Rwanda, Burundi, and Tanzania. The southern population is local in Zimbabwe, Botswana, Namibia, and South Africa (probably most common in southwestern Cape Province and the Transvaal highlands).

Measurements (Millimeters)
> **Wing:** males 165–175 (average of 4, 170.5), females 156–165 (average of 5, 161.0). **Exposed culmen:** males 38. 5–42 (average of 4, 40.6), females 35.5–40 (average of 5, 37.6) **Tail:** males 63.5–71 (average of 4, 67.3), females 63.5–71 (average of 5, 67.3)

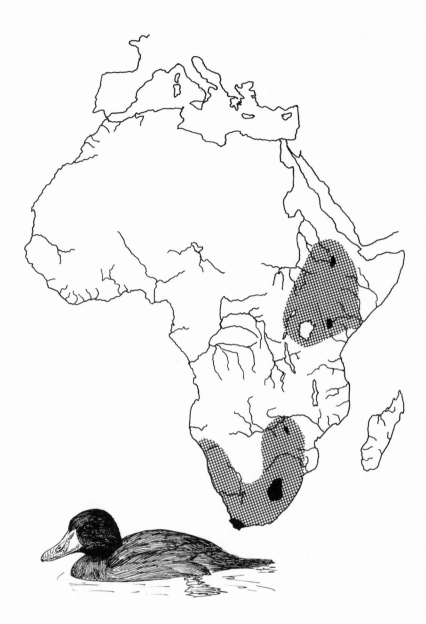

Figure 21. Distribution of the maccoa duck, showing the species' apparent residential range (cross-hatched) and areas of seemingly denser populations (inked).

(Clancey 1967). **Wing:** two males 165 and 173, one female 155; **culmen:** two males 40 and 42, one female 36; **tail:** two males 72 and 73, one female 66; **tarsus:** two males 34 and 34, one female 31 (Brown, Urban, and Newman 1982). **Maximum bill width:** males 22.5–24.8 (average of 8, 23.6) (Johnsgard 1967).

Weights (Grams)

Captives in England, one male 800, eleven females 580-720 (average 659.1) (Carbonell 1983). One wild male 820, three females 516–580 (average 554) (Brown, Urban, and Newman 1982). These highly inadequate weight samples suggest an average male-to-female mass ratio of about 1.3:1, which is a surprisingly unbalanced body mass that needs to be confirmed by larger sample sizes. If squared mean wing-length measurements are used as an alternative correlate of adult mass, the predicted male-to-female mass ratio is 1.12:1.

DESCRIPTION

Natal Down

The down is generally dark-grayish-brown above, with a less conspicuous dull white line below the eye than in *jamaicensis*, and dark cheek markings that are larger and less distinct (contra Delacour). The chin, throat, sides of neck, and underparts are off-white to grayish-white; the white dorsal patches are small and less evident than in *jamaicensis* (also contra Delacour) (Carbonell 1983). The downy tail feathers are dark brown, with stiffened shafts. A second generation of light-brown downy feathers follows the first natal down (Brown, Urban, and Newman 1982). The iris is brown; the bill, legs, and feet olive-gray to black.

Juvenal Plumage

The plumage generally resembles that of adult females but is duller and more uniformly colored (Delacour 1959). Feathers lack the subterminal blackish-brown spotting and incipient transverse barring of adults. Underparts are more rusty than in adult females, and the ventral feathers show more of the grayish-sepia feather bases (Clancey 1967). No doubt the tail feathers are distinctively notched terminally, the shafts with bare or still-downy tips.

First-Winter Plumage

The male's face is much like that of adult female, but with a broader diagonal blackish-brown cheek bar from the gape back to the side of neck; the hindneck and crown are more blackish-brown. Underparts are mostly dull grayish-sepia, the feathers tipped with grayish buff, and the undersurface often tinted somewhat vinaceous. Sides of breast, sides, and flanks are a dark chestnut, with buff and blackish vermiculations and granulations (Clancey 1967). From about seven months the males can be distinguished from females by their fading eye stripes and the brown feathers that replace an ash-brown mantle. The male's head later begins to turn black, sometimes in less than a year (Clark 1964; Carbonell 1983), sometimes as late as 20 to 21 months (Siegfried 1968), and the back color becomes a deeper brown. A lighter collar is retained until the head and mantle are adultlike, but the bill remains grayish-black for a longer period (Clark 1964; Siegfried 1968). Carbonell (1983) observed that two males that developed breeding plumages in their first year did so two months later than adult males, and their plumages were not quite so bright as in older birds, although their bills became bright blue.

Definitive Breeding Plumage, Male

This plumage is probably usually attained at about 20 to 23 months of age (sometimes when approaching 12 months), during the austral spring, before which males cannot be easily distinguished from females in the field. Entire head and upper neck are sooty black (with inconspicuous feathered "horns" on the crown); the upper foreneck and chin are dull. The remaining upperparts are chestnut, except for some of the long scapulars, which are basally olive-brown, and the middle back area, which is mostly dull brown. The lower throat, breast, body sides, and flanks are all chestnut, grading to grayish-brown on the abdomen. Undertail coverts are cinnamon. Underwing coverts are light brown, the axillars white. Wings are olive-brown dorsally, the tail blackish-brown. The iris is dark brown, the bill bright blue, with a darker nail; the legs and feet are grayish (Clancey 1967).

Definitive Nonbreeding Plumage, Male

The head is generally femalelike except for a darker crown, with a narrow but distinct subocular stripe, white cheeks, and throat. The breast is also mottled and femalelike, but with traces of chestnut present. Upperparts are mostly ashy-gray, with some chestnut. The bill is slaty gray. This plumage is gradually acquired during molt from February to May and held from May to July. By August the bill is again blue, and the plumage is becoming more black on the head and chestnut on the body (Siegfried 1968).

Definitive Breeding and Nonbreeding Plumages, Female

The crown and sides of the head are blackish-brown, with rusty fringes; the rest of upperparts are dark olive-brown, with tiny specks or bars of grayish buff and brown. The sides of the face are a dull buffy-white, heavily speckled with dark brown; there is a dark cheek stripe. The chin and upper throat are grayish-white; the rest of the underparts are pale sepia, the sides and flanks deep buff with sepia barring. Underwing coverts are pale grayish-brown, the axillars white; the upper wing is dark brown with buffy-gray granulations. The tail is blackish-brown. The iris is dark brown, the bill brownish to dusky, the legs and feet grayish. Presumably females have two body molts per year as do males, but this subject remains unstudied.

IDENTIFICATION (FG. 32B)

In the Hand

The combination of a bill length of 36 to 42 millimeters and a tail length of 63 to 75 millimeters separates this duck from other species of *Oxyura*. The bill is more distinctly swollen at the base than is true in the similar-appearing Peruvian ruddy duck, although not nearly so strongly as in the white-headed duck.

In the Field

This is the only species of stiff-tailed duck native to eastern and southern Africa; as such it can be readily identified by the long tail, diving behavior, and general chunky-shaped body profile alone.

ECOLOGY

Habitats and Densities

This species prefers shallow, nutrient-rich waters having extensive emergent vegetation, such as cattails *(Typha)* and reeds *(Phragmites)*. Waters that are not clogged with pondweed and floating vegetation may also be preferred. Bottoms that are muddy or silty and rich in larval dipterans provide ideal foraging opportunities. The species ranges from sea level in South Africa to as high as about 3000 meters in Ethiopia. Densities have not been calculated specifically, but Clark (1964) described two habitats and provided monthly counts throughout the year. One habitat (Modder East) consisted of a maximum of about 160 hectares (estimated from size description, assuming rectangular shape) and supported monthly averages of from 4 to 11 birds (average 7.7) over a four-year period. The other (Rondebult Sewerage Farm) contained an estimated 73 hectares of water and supported 9 to 33 adult, immature, and juvenile birds (average 22.2) over a two-year period. Among these were as many as 7 territorial males. This would suggest that in such favorable habitats from 5 to 30 adults may be expected per 100 hectares. The ratio of adult and subadult males among all birds counted in these two habitats was 48 percent for 611 birds, so that perhaps about 2.5 to 10 adult males per 100 hectares may be a reasonable estimate of male territorial density (Clark 1964).

Foods and Foraging

Foods are known to include the seeds of smartweed *(Polygonum)* and other plant species, the larvae and pupae of midges (Tendipedidae, Diptera), and various other invertebrates. Food is ingested at an estimated rate of 96 to 97 milligrams (dry weight) per dive (Brown, Urban, and Newman 1982). Durations of foraging dives by adults at the Wildfowl Trust were summarized in Table 6 and by ducklings in Table 7. Clancey (1967) reported foraging dives lasting 15 to 22 seconds. Clark (1964) similarly observed that foraging adult birds dive for 15 to 25 seconds, with intervening times on the surface of about 10 seconds, and that ducklings remain submerged in water about half a meter deep for 10 to 15 seconds (sample sizes not indicated).

Competitors, Predators, and Symbionts

No specific information on these topics exists. Several other ducks share habitats with the maccoas, such as the southern pochard *(Netta erythropthalma)* which probably consumes some of the same seeds of aquatic plants but is unlikely to be a direct competitor.

ANNUAL CYCLE

Movements and Migrations

In South Africa the birds are locally migratory in the Transvaal but are probably generally sedentary, depending on the permanent availability of suitable habitats. Clark (1964) reported some seasonal shuffling of populations in his Transvaal study areas, but some birds were present throughout the entire year.

Molts and Plumages

Molts have been described in general by Clark (1964) and Clancey (1967)—and by Siegfried (1968) for males specifically. Much of this information is still incomplete, but at least it is clear that males do have a definite femalelike nonbreeding plumage as do other *Oxyura* species. The molt generally occurs between May and July in South Africa, but in some (possibly immature) birds it may extend to October or even December, or during the major breeding season. The flight feathers are probably lost during this prebreeding period as well. There is also a full wing molt toward the end of February, during which time the tail feathers are also replaced; this molt is probably completed in all adult males by early to mid-May. This rather confusing pattern probably reflects both the long breeding season typical of South African maccoa ducks and the overlapping of molting cycles.

At the Wildfowl Trust, maccoa ducks regularly molt their wings and tails twice a year, the prebreeding molt beginning in January and the postbreeding molt starting in August. Both wing and tail feathers were molt rapidly, the wings somewhat before the tail and the tail feathers dropping more slowly, with new feathers growing in among the older ones. In Carbonell's study the remiges required about two weeks to regrow, and the rectrices two to three weeks. One female became flightless when her brood was only 17 days old but almost independent (Carbonell 1983).

Breeding Cycle

In South Africa the major breeding season generally extends from September to December, but some breeding occurs from June through the following April, with a subsidiary peak in February and March. The peak period in the Transvaal is from September to December and in the southwestern Cape from September to November. Some breeding occurs in east Africa during both the dry and the wet seasons, but too few records exist to provide any real pattern (Brown, Urban, and Newman 1982).

SOCIAL AND SEXUAL BEHAVIOR

Mating System

Siegfried (1976c) reported that maccoa duck males are strongly territorial, with displays geared to attracting and holding females and with a correspondingly promiscuous mating system. He observed as many as eight active nests within the territory of a single male and found individual males remaining sexually active and territorial for as long as four months. However, males outnumber females, and not all males can find even marginal territories, so some do not manage to breed at all. This species is probably among the most polygynous/promiscuous of all the *Oxyura*, and the reproductive success of individual males is apparently closely linked to their ability to hold territories and thus attract prenesting females.

Territoriality

Male territories are relatively large and may extend about 80 meters along the frontage of a stand of emergent vegetation (Siegfried 1976). Clark (1964) estimated that a territory may have a surface area of about 900 square meters or more and that breeding territories are bordered on one or more sides by reeds. In the course of a breeding season the territorial boundaries may change, as when more vigorous males become additionally established, so that where previously there were only two territories three or four may eventually be formed. Individual males were in territorial residence for from as short a period as five weeks to as long as three months in Clark's study area. Males unable to obtain territories gradually move away from the contested areas and gather with other nonbreeding birds in areas of open water.

At the Wildfowl Trust, Carbonell (1983) observed that males had established their territories by February or March. The males advertised their territories by performing the territorial vibration trumpet and by defending the territory against other males. Intruders were chased in a swimming-low-and-swift posture, diving, and actual fighting. During the major breeding season, at about the end of May, the two adult males shifted territories in at least two of the three years of study. The females evidently chose the male with whom to mate but did not necessarily nest in his territory. However, about a week prior to egg laying the female spent a considerable amount of time with the male, who displayed constantly toward her, dived with her, and showed her possible nesting sites (even in another male's territory), but was always ready to leave her if another receptive female came near. Thereafter the male stayed with the female most of the time until she began to incubate. At that point she became very aggressive, even toward this male.

Courtship and Pair-Bonding Behavior

Major studies of male advertising behavior have been performed by Siegfried and van der Merwe (1975) and by Carbonell (1983), and other relevant information has been provided by Johnsgard (1968a) and Clark (1964). Carbonell's description is most useful for the present summary, although the other contributions will be mentioned as appropriate.

The *ringing rush* (also "hurried flight" and "running flight") was performed as an apparent flight-intention movement and was also used by males to approach a female prior to displaying to her. The male lowered his tail, flattened his body and head feathers, stretched his neck forward, and rushed along the water while beating his wings (Fig. 22I). Contrary to Siegfried and van der Merwe (1975), this action only followed the head-high posture when it was performed as a flight-intention movement. In this posture the male approached the female until he was within about 1 meter, then assumed the rush (or "ski") posture, in which he swam with his tail flat on the water, the bill drawn against the breast, and the back of the body low in the water (Fig. 22J). Siegfried and van der Merwe reported that rushing is sometimes used aggressively, but this has not been observed by Johnsgard or Carbonell.

During the *swimming-low-and-swift* movement (or "stretch-swim" of Siegfried and van der Merwe), the bird lay flat on the water, with his bill partially submerged. The swimming-low-and-swift tactic was used to approach a female prior to displaying, to lead her to suitable nesting sites, to chase off intruders, and as a basic body position during courting, such as between other display actions. Siegfried and van der Merwe described a variant of this action, which they called the "leading swim," but not enough difference was observed at the Wildfowl Trust to warrant separation. The average length of display bout in the maccoa duck was 5 minutes; the maximum duration spent by a single male in uninterrupted display was 19 minutes. Males spent an estimated average of 13.5 percent of the daylight hours in display during the breeding season, a percentage second only to time spent by the white-headed duck (Table 3). The form and duration of the following display descriptions is based on the study of two males.

The *bill-dip, head-shake* display (Fig. 22E) represented 77.7 percent of all displays tallied. It lasted 0.2 to 0.4 second (average 0.27 second). This action is derived from a common comfort movement and begins in a swimming-low-and-swift posture. The bird dips his bill in the water and then pulls it out, shaking his head laterally. Unlike

Figure 22. Social-sexual behavior of the male maccoa duck: (A) vibrating trumpet; (B) territorial vibrating trumpet; (C) pumping (left) and choking (right) phases of sousing; (D) head-dip, head-flick; (E) bill-dip, head-shake; (F) swimming shake; (G) cheek rolling; (H) dab-preening, (I) ringing rush, and (J) rushing. (After sketches from cine films by Johnsgard 1968, Siegfried and van der Merwe 1975, and Carbonell 1983.)

head flicking, in the bill-dip, head-shake, the bill is not lifted above the vertical during the shake. In 45.9 percent of all displays recorded, the bill-dip, head-shake was the first ritualized action performed. It preceded all the other displays except for the wing shake, and was followed 53 percent of the time by cheek rolling. It was sometimes performed singly, but also in sequences of up to 17 dipping and shaking movements, with the male positioned laterally to the female.

The *cheek-rolling* (or "rolling-cheeks-on-back") display (Fig. 22G) represented 7.6 percent of the total displays tallied and lasted 0.6 to 1.1 seconds (average of 10, 0.86 second). Cheek rolling is a ritualized version of the corresponding comfort movement, except that the bird is positioned laterally to the female. He brings his head back over one side until it touches his back, then rolls it to the other side before returning it forward again. The movement typically followed the bill-dip, head-shake (89.5 percent of observations), swimming shake (5.3 percent), or sousing (5.3 percent) and was followed by either the bill-dip, head-shake (77.1 percent) or swimming shake (22.9 percent).

The *vibrating-trumpet* (or vibrating trumpet call) display (Fig. 22A) represented 6.3 percent of the total displays tallied and lasted 2.4 to 3.4 seconds (average of 12, 2.94 seconds). This is the most characteristic display of the maccoa duck. The bird starts from a swimming-low-and-swift posture, withdrawing his head and lifting it somewhat, with the bill open. The neck is inflated, and the neck feathers and "horns" on the crown are erected. The opened tail is then lifted vertically, the head is brought forward and downwards, and a belching call is uttered until the bill touches the water. The male positions himself laterally to the female, but as the action progresses he moves back in a semicircle around the female. Sometimes the tail is shaken at the end of the display. The vibrating trumpet is normally only directed toward females, but at times it may also be directed toward ducklings. In first-year males the belching sound is absent, as it also is in older birds at the start of the breeding season. In 90.5 percent of the cases this display was preceded by the bill-dip, head-shake; in 4.8 percent by the swimming shake or head shake. It was almost always followed by the bill-dip, head-shake but was once followed by sousing. In a variant form, the display was preceded by wing shaking; in this case the head and bill were already lifted from the water, and the back was hunched. At the end of this variant form the bill does not touch the water, and the neck is not stretched so far forward.

The *territorial vibrating-trumpet* (or "independent vibrating trumpet" of Siegfried and van der Merwe 1975) display (Fig. 22B) represents another variant form of the vibrating trumpet in that it is only performed by territorial males to mark their territories. It lasts 2.3 to 2.9 seconds (average of five, 2.47 seconds). It is very similar in form to the vibrating trumpet, but the action lacks the backward initial head movement; instead, the head is lifted vertically, then the neck is stretched forward and the bill is opened. The tail never reaches a vertical position, and the bill does not touch the water at the end of the display. Males will interrupt their other activities to perform this display periodically. Only once was a male observed to perform it in the middle of a courtship sequence. The average number of territorial vibrating-trumpet displays performed was 16 per hour, with a minimum interval between calls of 13 seconds and a maximum of 33 minutes. When the vibrating trumpet is performed by females it lacks vocalization and more closely resembles the territorial form of the display.

The *swimming-shake* (or general shake) display (Fig. 22F) represented 2.9 percent of the total displays tallied, and in one analyzed case lasted 1.1 seconds. As in other *Oxyura* species the movement closely resembles the corresponding comfort movement, in which shaking begins at the tail and proceeds forward, ending with a rotary movement of the head, and a final tail shake. This display was usually followed (54 percent of occasions) by the bill-dip, head-shake but also sometimes by cheek rolling or head dipping (15.4 percent each) or by vibrating trumpet or sousing (7.7 percent each). It was always preceded by bill dipping, head shaking or by cheek rolling (50 percent each).

Sousing (Fig. 22C) is a complex display that composed only 1.5 percent of the total displays tallied in courtship situations (also sometimes observed in alarm situations), but sousing sequences were the longest of all displays, lasting 4.0 to 19.9 seconds (average of six, 11.1 seconds). As in sousing by Argentine and Australian blue-billed ducks, the maccoa's sousing has two components. The initial phase, head pumping, is preceded by a head-high, tail-cock posture, the duration of which seemingly depends on the intensity. This is a vertical movement of the head and a forward movement of the cocked tail until it is diagonally over the back (Fig. 22C, left). The bill is then quickly lowered until it touches the water with each head-pumping movement, the number of pumps ranging from none to four (average of 11, 2.64). Siegfried and van der Merwe reported that head pumping was absent in most of the displays (80 percent) they analyzed. Following the head pump, the male reaches an extreme body position, with the neck fully swollen and the breast momentarily held out of the water. The choking (or "convulsive") phase then begins, as the bird drops back into the water and begins a series of rhythmical neck-up, head-down and neck-down, head-up movements, while his tail remains tilted above the back (Fig. 22C, right). The number of chokes varied from 1 to 3 (average of 12, 2.25) in uninterrupted sequence. The choking phase is followed by a head shake in the water, in which the male shakes his bill sideways in the water, while perhaps also expelling air, producing bubbles and splashes. However, no vocalizations are produced during sousing. Of eight sousing displays performed before a female, five were followed by bill dipping, head shaking, two by cheek rolling, and one by nothing. One of the sousings was preceded by the vibrating trumpet and another by a swimming shake.

The *head-dipping* display (Fig. 22D) represented 0.9 percent of the tallied display activities and lasted 0.4 to 0.6 second (average of four, 0.5 second). This display closely resembles the corresponding comfort movement; the low frequency of its occurrence may cause one to question if it is actually ritualized. Only once was it clearly directed toward a female, when it preceded a bill-dip, head-shake, and once it was performed at the end of a sequence of courtship displays. On three occasions it was followed by wing flapping.

Wing shaking represented 0.7 percent of the tallied display activities and lasted 0.4 second (two analyzed). It consists of a rapid shaking of the folded wings, which produces a rustling sound, as the tail remains flat on the water, the bill is pointed downwards, and the head lifted. This movement was always preceded by rushing ("skiing") and was always followed by a variant of the vibrating trumpet. The neck jerking mentioned by Johnsgard (1968a) does not appear to be typical.

Wing flapping represented 0.5 percent of the tallied display activities, and it is not clear whether it is ritualized or is only an associated comfort movement. However, it is always performed in the same manner in this situation—namely, following head dipping and before bill dipping, head shaking.

Head shaking represented 0.4 percent of the tallied display activities and is identical to the corresponding comfort movement. Here, unlike the bill-dip, head-shake, the bill does not touch the water.

Other authors have described various other movements as possible male displays, including dab-preening and tail shaking, but these actions either do not appear to be ritualized or are component parts of other displays mentioned above. Communal diving was observed on several times. Although Clark (1964) assumed it was of courtship significance, Siegfried and van der Merwe (1975) believed it to be unritualized. Judging from observations at the Wildfowl Trust, the latter interpretation appears to be true. However, simultaneous diving may sometimes occur among seemingly paired birds during the egg-laying phase of breeding.

Fertilization Behavior

Carbonell (1983) observed six full copulation sequences during her study, and Johnsgard has seen one. Siegfried and van der Merwe (1975) described copulatory behavior but did not indicate the number of observations on which their description was based. They believed that the vibrating trumpet call is an important, but not essential, stimulus for a female to accept a male, and she typically but not always responds to such a displaying male by performing water flicking before becoming prone on the water. Postcopulatory activity included bathing behavior by both sexes, consisting of head dipping and wing flapping, which the authors did not know whether or not to consider ritualized displays.

Johnsgard observed that a male performed two vibrating-trumpet displays, followed by cheek rolling and water flicking (bill dipping, head flicking) toward the female. The male then performed a long series of cheek-rolling movements as he approached the female, who did not overtly respond. He dived when about a meter away and soon came up directly beside the female. He then immediately mounted her, before she had become prone. No specific postcopulatory displays were performed, but normal bathing movements were performed by the female.

Carbonell observed that in three cases the male performed bill dipping, head flicking while approaching the female, and in another he performed a single vibrating trumpet call, to which the female responded first by pecking and then by going prone. On the other two occasions the male was actively displaying to one female when a second swam up and became prone. The postcopulatory behavior by the female consisted of preening in two cases and diving in three. In no cases did the female perform precopulatory head flicking or any specific postcopulatory displays, contrary to the account of Siegfried and van der Merwe (1975).

NESTING AND PARENTAL BEHAVIOR

Nests and Nest Building

Nests are usually constructed in reeds or waterside sedges; sometimes coot nests are used. Nests reportedly may even may be placed in willow cavities or under bushes in hollowed-out sand, although these situations must be regarded as highly exceptional. The nests are typically built of reed leaves that have been pulled down, crossed over, and bent to form a shallow basin about 8 centimeters deep and 20 centimeters wide. The nests are in contact with water, and the lowest part of the nest cup is about 8 to 22 centimeters above water level. Very little down is typically present, but during incubation the nearby reed stems may be pulled down above the nest to form a partial dome. Nests of separate females, presumably placed within a single male's territory, may be 5 to 40 meters apart (Clark 1964; Clancey 1967; Brown, Urban, and Newman 1982).

At the Wildfowl Trust 55 nests or attempted nests were found. All of the nests made among rushes, flag, or nettles (69.1 percent of the total) had partial roofs made of bent stems. Most of the nests were built less than a meter from the water; only a few were found more than a meter away. None of the nests were lined with down (Carbonell 1983). Measurements of these nests are summarized in Table 12. Males at the Wildfowl Trust spent a total of 3.2 percent of the day apparently exploring for possible nest sites with females, mainly during midday hours. A pair typically would be swimming, when the male would suddenly disappear in the vegetation. The female might follow or wait until he reappeared, at which time she would then enter the vegetation. On several occasions males were seen building rudimentary nest platforms. The pair would typically explore in several places, where the male would build platforms, until a site was finally chosen and the nest completed by the female.

Egg Laying and Incubation

During the first day of laying, and the day before it, female maccoa ducks at the Wildfowl Trust fed for a considerable amount estimated at nearly 90 percent of their available time. During the egg-laying period the females spent an estimated 34.3 percent of their time on the nest. The incubating females also spent an average of 91 percent of the daytime hours on the nest (Table 4), with bouts of sitting averaging 4 hours (5 observations), and 16 minutes off the nest (25 observations). In the case of two females from which the clutches were taken and renesting occurred, 20 and 25 days elapsed between the loss of the first clutch and the start of the second one. Some females laid at least three, and perhaps four, clutches within an overall laying period of five months (Carbonell 1983).

Eggs are laid at the rate of slightly less than one per day; one female laid two eggs in three days, another laid four eggs in five days, and a third laid six eggs in eight days (Clark 1964). The range of eggs in presumably single-female nests is 4 to 10, but normally consists of 5. Females at the Wildfowl Trust had clutches averaging 5.9 eggs and ranging from 3 to 10 (Table 14). Clutches of larger size (at least to 12) sometimes occur and presumably result from two or more females. Additionally, maccoa duck females sometimes lay their eggs in the nests of the Hottentot teal *(Anas punctata)* (Brown, Urban, and Newman 1982). Incubation is by the female alone and probably requires 25 to 27 days in the wild. It begins with the completion of the clutch; the process often fails apparently because of nest desertion, predation, or other unknown causes. The typical brood size at hatching is about three to four ducklings (Clark 1964). Incubation of eggs at the Wildfowl Trust required 23 to 26 days, and newly hatched ducklings averaged 56.3 grams (Table 14).

Hatching and Brood-Related Behavior

Little detailed information exists for wild birds. The brood is tended by the female alone; reports of possible male participation in brood care have not been verified. The family remains on open water during the day but may be brooded at night on the nest for the first few days following hatching. Females with broods may wander freely, and females respond to all males aggressively by gaping at them. Ducklings remain with their mothers for about five weeks, but the broodmates may continue to associate with one another for several weeks thereafter (Siegfried 1976c; Brown, Urban, and Newman 1982). At the Wildfowl Trust, Carbonell (1983) closely observed five females with broods of from two to seven ducklings. The females did not remove the eggshells after hatching; rather, they were left in the nest and crushed by the adult bird. At least two of the females returned with their broods the first night after hatching. All of them additionally brought their broods to land, or to old nest structures, for preening and resting until the young were about two weeks old. It required one female only two minutes to get her chicks initially out of the nest and into the water. Within a few minutes the ducklings began to dive, although they fed primarily by straining foods at the water surface. During the first two weeks of life the ducklings spent more time in foraging than in any other activity (Table 5). The maximum depth of the brooding pond was 2 meters. (The lengths of for-aging dives by maccoa ducklings is presented in Table 7.)

During the first week, the major activity of the ducklings was feeding (37 percent), followed closely by sleeping. The female's main activity was watching and remaining alert for disturbance. During the second week the ducklings spent more time feeding (54.1 per-cent) and less time resting, while the female's level of attentiveness also increased. During the third and following weeks, the amount of time the young spent in feeding diminished, and the resting periods increased. However, the female's alertness gradually decreased, and she spent more time in feeding and resting.

While standing or sitting, the female would brood the newly hatched young under her wing occasionally. It was not uncommon for young ducklings only a few days old to be swimming and diving alone and then climb up on land to preen and sleep. Eventually the female and the rest of the brood would come to this same resting site. The female's aggressiveness toward other birds gradually faded, as did her association with her brood. By the age of 14 days one male duckling was remaining almost entirely alone, and by 31 days he was completely independent. This same bird displayed toward his mother when 28 days old, by performing a display series (bill-dip, head-flick; cheek rolling; swimming shake; vibrating trumpet) and pecking at the female's neck. She responded with a bill-dip, head-flick. The young ducklings followed their mother while uttering a soft "cheep-cheep," while the mother constantly called with a low "gek-gek." With disturbance, she stretched her neck forward over the water to elicit a grouping response by the chicks. With further alarm the female called faster and louder and assumed a head-high posture, causing the ducklings to dive. Older ducklings (about six weeks old) would assume the head-high, tail-cock posture, with their breast sunk low in the water. The ducklings' primaries began to grow at 6 to 7 weeks, and the birds were fully fledged at 9 to 10 weeks. Activity budgets for adult females and ducklings at the Wild-fowl Trust are presented in Tables 4 and 5.

REPRODUCTIVE SUCCESS AND STATUS

There is no good information on reproductive success in this spe-cies in the wild. At the Wildfowl Trust 46.2 percent of 266 incu-bated eggs hatched, and 87 percent of these survived at least three to four weeks, representing an overall reproductive success rate of 40.2 percent. The sex ratio of these young birds was 1.1:1 (Carbon-ell 1983).

There is no good population data for wild birds, although Callaghan and Green (1993) have provided some rough estimates. The birds are probably most common in South Africa, especially in the southwestern Cape and parts of the Transvaal and Orange Free State. The maccoa is probably not regularly hunted, and thus its numbers are likely to be influenced by available breeding habitat. It has increased in some gold-mining areas, in association with habitat changes connected with these activities (Clancey 1967). A recent estimate of the world population is in excess of 5000 birds (Callaghan and Green 1993).

Two male Argentine blue-billed ducks courting a female; the more distant male is neck jerking.

CHAPTER TEN

Argentine Blue-Billed Duck

OXYURA VITTATA
(Philippi) 1860

Other Vernacular Names
> Argentine lake duck, Argentine ruddy duck, rusty lake duck; Argentinsche Schwarzkopfruderente, Sudamerikanische ruderente, Argentinische ruderente (German); erismature d'Argentine, erismature tachet, (French); pato malvasia mediano, pato pimpollo, pato rano de la Argentina, pato rano de pico delgado, pato tripoco, pato toro, pato zambullidor (Spanish); pato zambullidor chico (Argentina).

Range of Species (Fig. 23)
> Resident in the lowlands of Chile from Atacama south to at least Chilo, Island. Also breeds east of the Andes in Uruguay and Argentina, from San Juan and La Rioja south to Santa Cruz, more rarely to Tierra del Fuego, where it may breed in the northern part of Isla Grande (Gente Grande). Some seasonal movements of the eastern population occur in winter, to as far north as Paraguay and southern Brazil.

Measurements (Millimeters)
> **Wing;** male 142–168 (average of 3, 155.7), females 137–147 (average of 4, 143.8). **Tarsus;** males 29.7–35.5 (average of 4, 33.2), females 31.8–33.9 (average of 4, 32.9) (Carbonell 1983). **Cul-

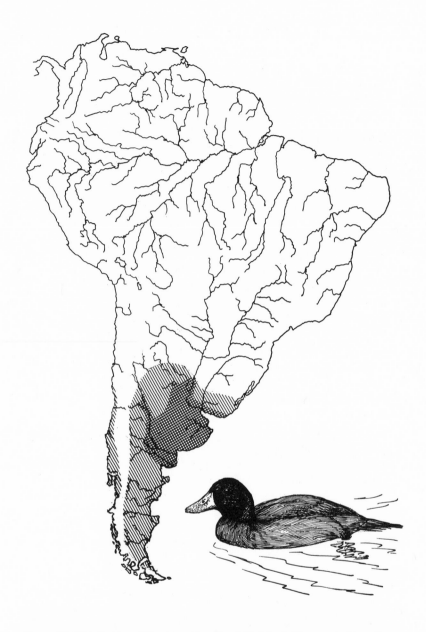

Figure 23. Distribution of the Argentine blue-billed duck: The species' breeding or residential range is shown hatched; the areas of denser populations are shown cross-hatched.

men: males 36–42 (average of 16, 38.7); **maximum bill width:** males 19.7–22.9 (average of 16, 20.6) (Johnsgard, pers obs.). **Maximum bill width** (sample size unstated); 19–23 (Phillips 1922–26). **Means of both sexes** (sample size unstated); wing 144, culmen 38.2, tarsus 33.7, bill width 22.1 (Johnson 1965).

Weights (Grams)

Wild males 623–850 (average of 3, 717.6), one female 623 (Humphrey et al. 1970). Captives in England, two immature (5 to 6 months) males 440 and 480, four females 510-700 (average 585.0) (Carbonell 1983). Single individuals, male 610, female 560 (Weller 1968a) These still inadequate samples suggest an average adult male weight of about 690 and an average female weight of about 590, representing an average male-to-female mass ratio of about 1.2:1. Based on squared mean wing-length measurements, the male-to-female mass ratio is also predicted as 1.2:1.

DESCRIPTION

Natal Down

The down is dark blackish-brown (darker than in *jamaicensis*), with no white except for a line below the eye and on the chin and sides of the neck; underparts are also whitish. There are no dorsal white markings on the back; downy tail feathers are dark brown and stiff; the bill, legs, and feet black.

Juvenal Plumage

This plumage is still undescribed in detail but is generally femalelike, although somewhat lighter overall. The juvenal tail feathers are notched terminally, their shafts with bare or sometimes downy tips, as in other juvenile stifftails (Weller 1968a).

First-Winter Plumage

"Immatures" are generally like the adult female but are lighter above, browner below, and their feather edges are brown or buffy rather than silver (Delacour 1959). Weller (1968a) stated that there is evidently a long-lived first nonnuptial (basic) plumage, as six young male specimens he observed had brown feathers replacing older brown feathers.

First Breeding Plumage, Male

First-year males at the Wildfowl Trust were somewhat duller in plumage than were the older birds, but by their second year they were not separable from older males (Carbonell 1983).

Adult Breeding Plumage, Male

The head and neck, except for foreneck, are black, with a white chin spot variably present. The remaining upperparts, plus breast, sides, and flanks, are bright chestnut. The abdomen and other underparts are silvery gray, with grayish feather bases sometimes showing through; there are never reddish tones. Wings are dark brown dorsally, more grayish and mottled white below, with white axillars; the tail is dark brown. The iris is dark hazel; the bill bright blue, with some black markings on the mandible, and the nail yellowish. The lower mandible is pinkish-yellow; the legs and feet olive-gray to olive-brown.

Adult Nonbreeding Plumage, Male

The head is generally grayish where it is normally black, but with some shades of chestnut also present. The remaining upperparts and underparts are mostly dark gray, with some scattered ruddy feathers. The bill is blackish (Carbonell 1983). Compared with females, the throat is whiter, and the body plumage is more rufous (Weller 1968a).

Adult Breeding Plumage, Female

The crown is brown, with indistinct black crossbars. A strongly contrasting whitish suborbital line extends from the bill to the nape, below which the cheeks are brown; the chin and upper throat in turn are a contrasting white. The underparts are silvery white except for the upper breast, which is barred with black. The upperparts are nearly black, with fine wavy bars (vermiculations) and frecklings of reddish or buff. Tail and wings are blackish, the latter with very fine freckling on the upper coverts and secondaries; the underside of the wing is as in male. The iris is hazel; the bill dark brown or blackish, with whitish edges and nail; the legs and feet bluish-gray.

Adult Nonbreeding Plumage, Female

This plumage is not yet specifically described but is certainly very much like the breeding plumage.

IDENTIFICATION (FIG. 33A)

In the Hand

The combination of a bill (culmen) length that is usually less than 40 millimeters (maximum 42 millimeters) and a maximum bill width of 22 millimeters separates this species from all other *Oxyura* except for the Australian blue-billed duck. This latter species usually has a bill length of at least 40 millimeters, although some overlap occurs in bill length and, indeed, in all other linear body measurements. Distinction between these two species must therefore rely on the more contrasting head patterns of females and immature or non-breeding Argentines, and, in breeding males, on the Argentine species' more silvery white abdomen (as opposed to a more mottled brown and white abdomen of the Australian).

In the Field

Except in southern Chile and adjoining southwestern Argentina, where the Peruvian ruddy duck may also occur, there should be no problem identifying this species, for it is likely to be the only waterfowl with a lengthened tail. In areas of overlap with Peruvian ruddy ducks, the Argentine blue-billed duck appears significantly smaller, and females (as well as nonbreeding or immature males) have much more strongly contrasting facial markings. Argentine blue-billed ducks are more likely to occur on lowland marshes and rarely extend to alpine lakes as does the Peruvian ruddy. In some areas the Argentine blue-billed duck comes into contact with the masked duck; here field identification is easier, based on the size differences of the two and the distinctive appearance of the masked duck, with its rather short and relatively straight-in-profile bill (rather than a distinctly concave bill) and a tendency for overall plumage spotting, rather than vertical barring, on the flanks and upperpart plumage of both sexes.

ECOLOGY

Habitats and Densities

In Argentina this species occurs on large lakes and semiopen marshes with large pools. Deep, open roadside marshes are also used, especially if they adjoin large marshes and have interconnecting waterways. Pools that are covered by duckweeds (*Azolla* and *Lemna*) are also used during the breeding season (Weller 1967b).

Foods and Foraging

There is no direct information, but the birds are known to feed on duckweeds at the surface and to dive in waters having muddy bottoms, both in captivity at the Wildfowl Trust and in the wild. These behaviors suggest that they feed on the same mixture of plant seeds and aquatic invertebrates such as dipteran (especially midge) larvae that are typical in the diet of other *Oxyura* species. Diving times for adults at the Wildfowl Trust are presented in Table 6 and for ducklings in Table 7. General activity budgets for adult males, including time spent foraging, are shown in Table 3; those for females are shown in Table 4.

Competitors, Predators, and Symbionts

Argentine blue-billed ducks occur in areas also occupied by black-headed ducks. There are no records of parasitism by this species in the wild, but there are numerous records of this occurring at the Wildfowl Trust (Table 17). There is also limited sympatry with the Peruvian ruddy duck in Chile (Johnson 1965) and in Argentina (Scott and Carbonell 1986; Casos 1992), where Carbonell has seen the Argentine species as high as 1200 meters above sea level and the Peruvian ruddy duck to about 900 meters. No definite hybrids between these species have been found. Probably few avian predators affect adult or older juvenile blue-billed ducks because of their excellent diving abilities, but the ducklings no doubt are eaten by many predators.

ANNUAL CYCLE

Movements and Migrations

This species is apparently residential through its range, although certainly in southern Argentina and Tierra del Fuego there may well be some seasonal movements. As noted in Chap. 3, Todd (1979) reported the dispersal of some birds to Deception Island, about 1000 kilometers south of Tierra del Fuego and directly north of the Antarctic Peninusula.

Molts and Plumages

The molting cycles have not yet been studied in the wild, although Weller (1968a) provided some comments. He believed that young male birds first acquire their breeding (alternate) plumage from September to October, and a late winter or early spring (August to October) body and tail molt probably occurs in both sexes. The nonbreeding (basic) plumage may develop between late January and May. At the Wildfowl Trust, captive Argentine blue-billed ducks molted twice a year. The prebreeding molt began in late December, when the males' bills turned bright blue, the head attained its black feathers, and the body plumage became bright chestnut. By mid-January the males were all in full breeding plumage and had just become flightless. Soon after the remiges were lost, they also dropped their rectrices. Sometimes the tail feathers were all dropped within a few days, with the birds then being tailless for a short time, but in most cases the new tail feathers grew in among the older ones. The postbreeding molt occurred at the end of September, when the birds underwent a second flightless and tailless period. The remiges were lost first, and while the bird was flightless, the rectrices were also shed. While these feathers were growing back, the body and head acquired the dull coloration of the nonbreeding plumage, and the bill turned black. Both the primaries and the rectrices required approximately two to three weeks to regrow (Carbonell 1983).

Breeding Cycle

Little information exists on the breeding phenology of wild birds. Nesting in Chile (Aconcaqua) reportedly occurs twice a year, but actual nesting dates were not provided (Johnson 1965). In central Argentina the nesting season is reportedly from mid-October to early January (Weller 1967b). At the Wildfowl Trust this was the first of the stifftail species to acquire its breeding plumages seasonally, but males did not start displaying until the end of January. The females began to lay during May, and the last eggs were laid in late August to early September (Carbonell 1983).

SOCIAL AND SEXUAL BEHAVIOR

Mating System

No information exists for wild birds. At the Wildfowl Trust the pair bond was stronger than in other *Oxyura* species. One pair remained together throughout the three-year study period, but two other females did not remain paired after breeding. It was impossible to determine whether they paired with the same males again the following year, but at least temporary monogamous bonds were established. However, both of the males involved in these pairings attempted to pair with an extra female as soon as their mates began to incubate. Generally the male would initially remain near its mate's nest, displaying intensively, but after a few days of incubation would move away in search of the unpaired female.

Territoriality

Males at the Wildfowl Trust did not defend any definite territory except for the area immediately around the female and the site chosen by the female for nesting. The latter was defended until the female began to incubate. Most of the chasing away of intruders was done by the female; males chased other females away from the nest area only when courtship display occurred (Carbonell 1983).

Courtship and Pair-Bonding Behavior

The only prior detailed description of displays in this species is that of Johnsgard and Nordeen (1981), which was also based on observations of captive birds at the Wildfowl Trust. Weller (1967b) had earlier described some of these same displays in wild birds. The form and duration of the following display descriptions is based on Carbonell's study of three males.

When separated from their females, males swim very fast with the back of the body lowered in the water, breast above the surface, and bill against the breast. This display has been called the "rush posture" in the North American ruddy duck, "skiing" in the maccoa duck, and, in less extreme form, "motor-boating" in the Australian blue-billed duck. This posture was never as exaggerated in the Argentine species as in the maccoa duck or North American ruddy duck, more closely resembling the motor-boating display of the Australian species.

"Flotilla swimming" has been observed throughout the breeding period and involves both sexes, paired as well as unpaired individuals. The birds swim faster than normal, from one end of the pond to the other, with their tails flat on the water and uttering a "prrr-prrr" noise. The males constantly try to move closer to one of the females and sometimes perform wing-shaking displays toward one before moving away toward another female. Such bouts of swimming usually last about 6 minutes, but at times they last as long as

17 minutes (Carbonell 1983). Somewhat similar flotilla swimming occurs in the white-headed duck, but its possible role in courtship and pair bonding of both species is still unknown.

Male Argentine blue-billed ducks do not seem to position themselves in any special way relative to the females, except when performing sousing and dab-preening. All of male displays except for sousing and wing shaking have an associated "prr-prr" sound, presumably of vocal origin. The average duration of male display bouts was 3 minutes; the maximum observed duration of display by a single male without interruption was 16 minutes. During the breeding season males spent about 5.5 percent of their daytime activities in sexual display (Table 3). This is the lowest percentage of daytime display of any species analyzed by Carbonell. Perhaps the apparently low incidence of display reflects the monogamous pair bonding, rather than polygynous mating, that is seemingly typical of the species.

Neck jerking (Fig. 24A) is the most common male sexual display of the species, comproing 58.3 percent of all displays tallied by Carbonell. In this display the male lifts his head very quickly, bringing it backwards slightly, and pulling the breast well out of the water. Then the head is lowered much more slowly, with a slight movement forward, while the tail remains on the water. Neck jerking somewhat resembles an exaggerated form of the head-high posture and is performed independently of, or together with, other displays. The mean duration of 54 displays was 0.35 second (range 0.2 to 0.5). It is always directed toward the female. Up to 29 jerks may be performed without interruption, with 5.6 being the average. Similar results (average of 5.5 in a range from 1 to 13) were obtained by Johnsgard and Nordeen (1981).

The *bill-dipping* display (Fig. 24F) is a slightly ritualized version of the corresponding comfort movement; it consists of dipping the bill and then lifting it back to its normal position. No sideways head shaking was observed by Carbonell, although Johnsgard and Nordeen observed this component. Bill dipping represented 21.6 percent of all the displays tallied; 16 displays lasted an average of 0.35 second (range 0.2 to 0.4 second). This action always preceded the double head flick prior to dab-preening, but in one case the bill dip was followed directly by dab-preening. In such cases the bill dip was modified in that the bill was not oriented directly in front of the body but rather was directed toward the female.

In the *cheek-rolling* (also head-rolling, rolling-cheeks-on-back, cheek-rubbing) display, which represented 4.4 percent of all the displays tallied, the male brings his head back until it touches his back and rolls it from side to side (usually) before bringing it back to the normal position. It differs scarcely if at all from the corresponding comfort movement; 50 displays lasted an average of 0.42 second (range 0.2 to 0.6 second).

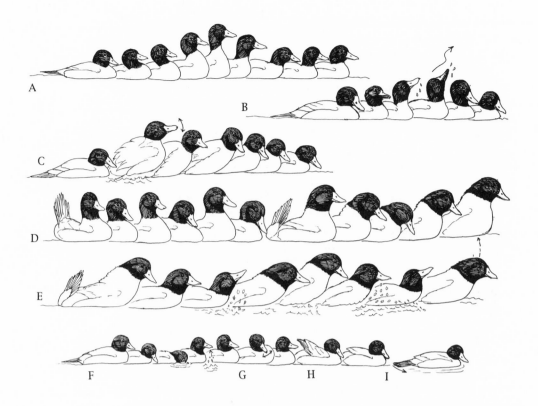

Figure 24. Social-sexual behavior of the male Argentine blue-billed duck: (A) neck jerking, (B) head flicking, (C) swimming shake, (D) pumping phase of sousing, (E) choking phase of sousing, (F) bill dipping, (G) dab-preening, (H) wing shaking, and (I) tail shaking. (After sketches from cine films by Carbonell 1983 and Johnsgard and Nordeen 1981.)

The *head-dipping* (or dipping; also "splash bathing" of Johnsgard and Nordeen 1981) action, which represented 4.0 percent of all displays tallied, is a ritualized version of the corresponding comfort movement. The male dips his head forward until it is beneath the water, then quickly lifts it so that water splashes over the back. The display ends with a tail shake. It was always performed in the presence of a female. Nine displays lasted an average of 0.57 second per dipping movement (range 0.4 to 0.7 second).

Sousing (the choking display of Weller 1967b) is the most complex and prolonged display of the species; six sequences analyzed by Carbonell lasted an average of 10.5 seconds (range 5.6 to 12.4 seconds). However, sousing composed only 3.2 percent of all displays tallied. Of the six sequences analyzed, the average number of pumping movements was 6.2, and the average number of chokes per sequence was 9.5. Only those sousings directed toward females were considered in calculating frequency of occurrence; sousing sometimes is also provoked by alarm stimuli. The sousing sequence consists of two components, the head pump (Fig. 24D) and the choking (Fig. 24E). The bird first assumes a head-high posture, followed by the head-high, tail-cock, while uttering a "prr-prr" call; the head pump then follows, consisting of vertical head movements, in which the bill is brought down to touch the breast. With each pump the neck is progressively inflated, and the tail moves increasingly toward the back. The number of head-pump movements varies from 4 to 10 (average of 28, 6.8). Gradually the pumping becomes more of a rocking movement, until the male's body is in an almost vertical position, with his breast out of the water and his neck feathers fully erected. Then the bird drops suddenly into the water, producing a splash, and begins the choking phase. In this part of the sequence the male utters a repeated honking call, which is synchronized with the convulsive movements of alternated neck-up, head-down and neck-down, head-up postures. The numbers of chokes performed in sequence varies greatly, ranging from 3 to 17 (average of 33, 9.0) perhaps depending on the intensity of the initiating stimulus, but is most often 9 to 10. Although Johnsgard and Nordeen only observed the display performed directly in front of females, about half of those that Carbonell observed were performed in the absence of females, usually when the male had been alarmed by a human or predator. Sousing displays that were performed toward females were sometimes followed by neck jerking.

Head shaking, which composed 2.3 percent of the tallied displays, is a slightly ritualized version of the normal comfort movement. The bill is held below the horizontal and shaken laterally; thus head shaking differs from the head flick, in which the head is shaken with an upward movement. Its usual duration of a head shake was 0.2 to 0.3 second (average of 3, 0.26 second).

Head flicking (Fig. 24B) composed 1.7 percent of the tallied displays and lasted about 0.2 to 0.3 second (average of 15, 0.29 second). It consists of a lateral head shake as the bill is tilted upwards. Carbonell reported seeing it only as an independent activity, not as part of the general swimming shake as observed by Johnsgard and Nordeen.

Dab-preening (Fig. 24G) composed 1.5 percent of the tallied displays and lasted about 0.6 second. It is performed laterally to the female, with the bill brought down against the breast and moved laterally, as if nibbling the side of the breast facing the female. It was always observed to be followed by a wing shake and tail shake, the three forming a fixed sequential pattern dab-preen, wing shake, tail shake. It also usually followed bill dipping and once followed head jerking. It was observed that following clutch loss this display was always performed by a male toward his mate, typically during the first or second day after her clutch had been removed.

The *wing shake* (Fig. 24H) composed 1.6 percent of the tallied displays and lasted about 0.4 second. It is a ritualized version of the corresponding comfort movement and consists of a rapid shake of the wing feathers as they are brought up to the center of the back. An associated stridulation sound is produced. The majority of observed wing shakes were performed immediately after dab-preening, and the rest were seen in a noncourtship context, during group swimming (flotilla swimming).

The *swimming shake* (or general shake, body shake) (Fig. 24C) composed 1.1 percent of the displays tallied and lasted 0.1 to 1.0 second (average of 8, 0.71 second). It is a ritualized version of the general body shake, which begins with a tail shake. The shaking then proceeds forward until it reaches the head, which is rotated along its long axis. As this occurs, the bird paddles its feet, raising its body from the water, then finally sinks back into the water and shakes its tail again. This sequence of movements is contrary to an earlier and less acurate description by Johnsgard and Nordeen.

Fertilization Behavior

Copulation or attempted copulatory behavior was observed by Carbonell (1983) on seven occasions. In all cases both sexes were initially diving. In four cases the male stopped diving and began to perform neck jerks (maximum of six seen) before mounting the female. On another occasion he may have performed head pumping, but the light was inadequate to be certain. On the two remaining occasions the precopulatory behavior was not observed. The female performed neck jerking before one of the copulations. In the other cases she had been diving and, without any special signaling, became prone and allowed the male to mount her. Diving or preening was performed by the female following copulation; in most cases the male simply preened afterwards.

NESTING AND PARENTAL BEHAVIOR

Nests and Nest building

Only a few nests have been described from the wild (Weller 1967b; Johnson 1965), and these do not differ significantly from the nests constructed by captive females at the Wildfowl Trust. Of the 29 nests or attempted nests observed there, 55.2 percent were found inside nesting boxes, and 44.8 percent were located among shoreline vegetation such as rushes. Those found in boxes were made from old nesting materials left inside and new materials gathered from around the box. The nests found among rushes were made from broken pieces of this plant; the nearby stems had been bent down to create a partial roof overhead. None of the nests contained any down (Carbonell 1983). Measurements of these nests are summarized in Table 12.

Males spent 2.2 percent and females 1.3 percent of their time exploring for nest sites. This was done mainly in the evenings. Typically, a pair would be feeding or swimming about in the pond, and the male would at some point disappear into the vegetation while the female stayed in the open water. After a while he would appear and the female would then enter the vegetation. The male remained nearby, swimming and performing head jerks in a seemingly nervous way. Several times the males were seen passing material and building a very rudimentary platform at such places. A pair might explore several sites at which the males would build platforms, until a nesting site was seemingly chosen. The female would then build the rest of the nest by herself (Carbonell 1983).

Egg Laying and Incubation

No detailed information exists for wild birds. At the Wildfowl Trust females spent an estimated 87.3 percent of the daylight hours feeding during the day or two prior to laying and on the first day of laying. During the egg-laying period they spent about 34 percent of their time on the nest; during incubation they spent 93.3 percent of their time there (Table 4). The bouts of sitting on the nest averaged 4 hours, as compared with 16 minutes off the nest (25 observations). The average interval between the end of incubating one clutch (when it was removed) and the onset of laying a replacement clutch was 23.5 days (6 observations, range 17 to 33 days).

Clutch sizes in the wild typically are of 3 to 5 eggs, although as many as 12 eggs have been found in a single nest (Johnson 1965; Weller 1967b). The females at the Wildfowl Trust had average clutches of 4.3 eggs, ranging from 3 to 10. The incubation period has not been estimated in the wild, but at the Wildfowl Trust it covered 23 to 24 days; newly hatched ducklings averaged 48.4 grams (Table 14).

Hatching and Brood-Related Behavior

There is information found on wild birds. Certainly the young are tended only by females, and like other stifftails, the ducklings probably become independent very rapidly. Among ducklings reared at the Wildfowl Trust, their primaries started to grow at five to six weeks; fledging occurred at eight to nine weeks (Carbonell 1983).

REPRODUCTIVE SUCCESS AND STATUS

There is no information on reproductive success in wild birds. Of 93 eggs incubated at the Wildfowl Trust, 50.3 percent hatched, and 80.8 percent of those survived to at least three to four weeks, for an overall reproductive success of 40.9 percent. The sex ratio (male to female) of these ducklings was 1.1:1 (Carbonell 1983). Apparently very few of these ducks are shot by hunters; Weller (1968b) found only two among a sample of 263 hunter-killed ducks.

Two male Australian blue-billed ducks courting a female; the nearer male is sousing.

Australian Blue-Billed Duck

OXYURA AUSTRALIS
Gould 1836

Other Vernacular Names
> Australian ruddy duck, blue-billed duck, diving duck, little musk duck, spine-tail, spiny-tailed duck, stifftail, stiff-tailed duck; Australische Ruderente (German); erismature d'Australie (French); pato pico azul (Spanish).

Range of Species (Fig. 25)
> Resident, with some local dispersion or seasonal migration, in southwestern Australia from Roebuck Bay to Cape Pasley, and from eastern South Australia (Spencer Gulf and Lake Eyre region) east through Victoria, the Murray-Darling basin of New South Wales, to southeastern Queensland (Hervey Bay). Also resident in Tasmania.

Measurements (Millimeters)
> **Wing:** males 150–173 (average of 222, 160), females 142–163 (average of 121, 153). **Culmen:** males 37–48 (average of 231, 41), females 32–47 (average of 122, 41) (Frith 1967). **Tail:** males 64–71 (average of 13, 66.5), females 57–68 (average of 7, 64.6). **Tarsus:** males 35.4–39.2 (average of 16, 37.3), females 35.5–37.9 (average of 8, 37.1) (Marchant and Higgins 1990).

Weights (Grams)
> Adult males 610–965 (average of 241, 812); adult females 476–1,300 (average of 140, 852) (Frith 1967). Adult females, during

Figure 25. Distribution of the Australian blue-billed duck: The species' breeding or residential range is shown hatched; the areas of denser populations are shown cross-hatched.

prebreeding period, average of 47, 798; during prelaying period, average of 11, 1,010; during laying period, average of 10, 1,167; during nonbreeding season (December to May), average of 26, 624 (Briggs 1988). The mean weight statistics provided by Frith would suggest an average male-to-female mass ratio of 0.95:1, representing a case of slightly reversed sexual dimorphism that would be unique among all *Oxyura*. However, female wing, tail, and tarsus measurements all average smaller than those of males, and the nonbreeding season and prelaying female weights established by Briggs also indicate much lower mean female weights during these periods than Frith's data would suggest. If Frith's average weight for males (812) and Briggs's overall mean weight for nonbreeding and prebreeding females (736) are used, the estimated male-to-female mass ratio is 1.1:1. If the square of the mean wing length as an indirect mass estimate for each sex is used, the predicted male-to-female mass ratio is also 1.1:1, which is much more in line with corresponding mass ratios of closely related *Oxyura* species, such as that of the Argentine blue-billed duck.

DESCRIPTION

Natal Down

The down is dark brown above, with obscure facial markings and small, inconspicuous lighter-grayish patches on the upper back and in the pelvic region. The throat and abdomen are grayish-white. Downy tail feathers are dark brown and stiff. The iris is blackish-brown; the bill grayish-black, with a paler base; the legs and feet dark gray to greenish-flesh, with darker webs.

Juvenal Plumage

This plumage is very similar to that of the adult female, but body feathers are narrower and perhaps slightly paler. Tail feathers have notched, bare tips. Softpart colors are as in older females. There is a complete postjuvenal molt (Marchant and Higgins 1990).

First-Winter Plumage

This plumage is not yet adequately described but is generally femalelike. Delacour (1959) stated that "immatures" are less clearly barred than adult females.

Definitive Breeding Plumage, Male

Judging from observations at the Wildfowl Trust, this plumage, and associated sexual maturity, is evidently attained in the first year of life. The head and neck are black, with a brownish tinge to chin and throat. The upperparts are mostly reddish brown; the longest scapulars, the rump and uppertail coverts are blackish-brown with freckled cream bars. Other feathers have buff to reddish-brown speckling or feather edging; the feathers of the upper breast, axillars, and flanks are reddish-brown; the lower breast and underparts to tail coverts are mottled brown and whitish, the tail coverts dark brown, with conspicuous whitish feather tips. The wing is grayish-brown dorsally, lighter gray below; the tail is dark brown. The iris is reddish-brown to blackish-brown. The bill is bright blue basally, becoming paler toward the tip and around the nostrils; the edges of upper mandible are whitish, the lower mandible cream-colored. Legs and feet are gray (Marchant and Higgins 1990; Phillips 1922-26).

Definitive Nonbreeding Plumage, Male

This plumage is very similar to that of the adult female, but the head is generally darker, the flanks sometimes have chestnut present, and often some rufous feathers are present on the upperparts.

Definitive Breeding and Nonbreeding Plumages, Female

No marked seasonal changes are present. Crown, face, and hindneck are blackish-brown, barred or freckled with yellow-brown. The chin and upper throat are white; sides of the neck, face, and especially the ear coverts are irregularly barred with dark brown, but no definite cheek stripe is present. Upperparts are mostly blackish-brown, with white, buff, or light brown irregular barring. Upper breast and flanks are irregularly barred dark brown and yellow-brown; the rest of the underparts are mostly white, mottled with dark gray. The wing is brown dorsally, underwing coverts are grayish-brown, the tail dark brown. The iris is brown; the bill blackish-brown with lighter edges; the legs and feet gray (Marchant and Higgins 1990).

IDENTIFICATION (FIG. 33B)

In the Hand

The combination of a bill (culmen) length that is usually between 40 and 42 millimeters and a maximum bill width of 20 to 22 millimeters separates this species from all other *Oxyura* except the Argentine blue-billed duck. Females, and males other than those in breeding plumage, of this species differ from the Argentine species in that they lack a strongly patterned face. Breeding males are mottled with pale brown and white on the abdomen, rather than being silvery white in this area.

In the Field

This and the musk duck are the only two Australian stifftails and can be easily distinguished at any distance by their great size difference alone, the blue-bill being roughly coot-sized and the musk duck almost goose-sized. Additionally, some chestnut feathering is likely to be present on adult male blue-billed ducks, regardless of their seasonal plumage stage; there is no trace of the large gular wattles that male musk ducks conspicuously exhibit.

ECOLOGY

Habitats and Densities

During winter, blue-billed ducks congregate on large, clear lakes, sometimes in the thousands of birds. Johnsgard (1965b) counted over 1600 birds along a measured mile of shoreline at Lake Kangaroo, Victoria, in mid-July, and this was only a small part of the lake's total population at that time. With the coming of spring, or at least with a period of extended rainfall, the birds disperse into deep and heavily vegetated swamps and marshes. These are often lined with cattails *(Typha)*, but the birds also at times occur in lignum *(Muehlenbeckia)* swamps or, in coastal areas, dense and swampy tea-tree *(Melaleuca)* thickets. They occasionally also breed on river oxbows (billabongs), in flood-filled depressions on the plains, or along river frontages (Frith 1967).

Foods and Foraging

At Barrenbox Swamp, New South Wales, these birds made extensive use of permanent cattail or "cumbungi" *(Typha)* swamps for foraging, frequent use of aquatic plants in shallow water, and rare use of the rush and sedge shoreline zone for foraging. Like musk ducks, the birds always fed in the deepest water and in the densest part of the swamp. In terms of estimated volumes of foods consumed by analyzing the stomach contents (which overestimates the amounts of harder materials) of 593 birds, 54.2 percent came from plant sources and 45.8 percent from animal materials. Of the plants, 23.8 percent were deep-water plants, 13.6 were edge plants, 9.6 were floating plants, and the remainder were littoral and dryland species. The most important plant groups represented were the coontails (Ceratophyllaceae, 13.4 percent of total volume), the smartweeds (Polygonaceae, 11.2 percent), and cattails (Typhaceae, 5.3 percent). There was little indication of seasonal changes in foods taken; the variations observed seemed to reflect mainly the cycle of abundance of various plant materials. Over 20 families of plants were represented, but no single plant family clearly dominated the diet. Of the animal materials consumed, 42.6 percent of the volume was of insects, and crustaceans made up most of the small remaining amounts. Most of the insects eaten were larval midges (Chironomidae) (23.5 percent of total food volume), with larval caddis flies (Trichoptera) in secondary position (9.7 percent) (Frith, Braithwaite, and McKean 1969).

Some foraging-dive durations observed for this species are presented in Table 6 Frith (1967) similarly reported that most foraging dives last 10 to 15 seconds, but some occasionally last as long as 30 seconds, with the birds diving in waters up to about 3 meters deep.

Competitors, Predators, and Symbionts

Probably the musk duck is the nearest foraging competitor of this species. In areas where they occur together they forage in the same situations—namely, deep waters of densely vegetated swamps. Thus at Barrenbox Swamp they foraged together but, as compared with blue-billed ducks, musk dusks fed relatively little on dryland plants, edge plants, cosmopolitan plants, or littoral-zone plants. Additionally, musk ducks ate far more crustaceans, and these were larger forms, rather than consuming the microscopic-sized cladocerans and ostracods that were taken by the blue-billed duck. Contrariwise, blue-billed ducks specialized in chironomid larvae, which were rarely eaten by musk ducks (Frith, Braithwaite, and McKean 1969).

Predators are largely unknown, but water rats (*Hydromys chrysogaster*) may be responsible for many egg losses (Marchant and Higgins 1990).

ANNUAL CYCLE

Movements and Migrations

The movements of the blue-billed duck are basically regular, so that local populations tend to fluctuate annually. In areas such as Lake Wendouree, Ballarat (Victoria), the birds regularly arrive for breeding and depart afterwards, with numbers varying from year to year, depending on water conditions in eastern Australia generally. A similar pattern occurs at Lake Barrenbox, New South Wales. On the Murray River, and in the larger swamps and lakes of the Murray River basin (such as Lake Kangaroo), the birds concentrate and molt. From there they largely disperse again back into the swamps of New South Wales for breeding, although some remain to breed locally in Victoria (Frith 1967).

Molts and Plumages

Little information exists on wild birds. Marchant and Higgins (1990) reported that all juveniles molt at 2 to 3 months and that captive males assume their first breeding plumage at about 10 months. Adults have two complete wing molts per cycle, with accompanying "normal" body molts. The first complete molt occurs at 6 months of age, and thereafter adults molt at about 6-month intervals. Adult males begin their prebreeding molt in July and their postbreeding molt in December. Adult females begin their prebreeding molt in June and July and their postbreeding molt in January. The presence of mature males may delay the assumption and also shorten the duration of the breeding plumage in young males. Some captive juveniles kept with dominant, "permanently colored" adults assumed only 20 percent of their breeding plumage as yearlings and required more than two years to attain their full breeding plumage. However, an isolated group of captive juvenile males assumed their first adult breeding plumage at 14 months of age. By 16 months the birds were in about one-third nonbreeding plumage, and by midwinter half were of the nonbreeding type. Their prebreeding molt began in October, and by mid-November their full breeding plumage and bright bill color had been regained.

Breeding Cycle

Breeding data were obtained by Braithwaite and Frith (1969) from 269 adult males and 80 adult females collected at Barrenbox Swamp, New South Wales, over a five-year period. In spite of the disproportionate number of males in the sample, the authors judged that the actual sex ratio was near unity. Courtship was observed from August to November, and gonad examinations indicated that the breeding cycle occurs regularly in the September to November period. However, if conditions are favorable, some breeding may occur in every month of the year, since it was virtually continued from October of 1963 to February of 1965. The male testis cycle peaked in spring or early summer (October to December); shortly after peak activity there was an exodus of birds and little gonadal activity. The females showed an annual pattern of gonadal activity corresponding to that of the males, with most birds having enlarged follicles from August to October or November. Unseasonal breeding coincided with periods of relatively stable autumn water levels, for reasons that were unclear but may have been related to the supplies of their principal animal foods, larval midges and caddis flies.

SOCIAL AND SEXUAL BEHAVIOR

Mating System

This species was judged recently by Marchant and Higgins (1990) to have short-term pair bonds, with probable serial polygyny prevailing. At least in captivity, a single male can dominate at least three females, whose clutches may overlap temporally. Such a dominant male defends all these females, repels competitors, and suppresses the courtship behavior of other males. This situation is probably most similar to that of the North American ruddy duck and the white-headed duck.

Territoriality

Males are apparently only sporadically territorial. Unlike maccoa ducks, the male does not defend a defined area around his nesting female or females and apparently does not try to defend a defined area around the nest. Rather, a mobile territory seemingly exists, in that males will repel others that approach their own females, in the same way that females have a portable territory surrounding ther broods (Marchant and Higgins 1990).

Courtship and Pair-Bonding Behavior

Various displays of this species have been described, including accounts by Wheeler (1953) and Scott (1958). The description by Johnsgard (1966) was the first serious attempt to describe all the male courtship displays but was based on short-term observations of wild birds and was admittedly incomplete. It has recently been supplemented by additional information in Marchant and Higgins

(1990), which was based mostly on captive birds studied by Rushton and Fullagar. However, a more detailed account of sexual behavior in the species is still needed.

The "display flight" of males was observed by Johnsgard only twice and appeared to correspond exactly to the comparable ringing rush of the North American ruddy duck. It consists of a short, "buzzing" flight by the male slightly above the water surface and sometimes serves to bring the male into closer proximity to a female. However, it may also serve to take the bird away from the group or from an opponent. In one of the two cases noted by Johnsgard the male terminated the flight with bill dipping and wing flapping. The flight may transport the bird 10 meters or more and last one to two seconds. Under favorable conditions the splashing noises accompanying this display can be heard more than a kilometer away (Marchant and Higgins 1990).

The swimming-low-and-flat tactic (Marchant and Higgins, 1990) may also be used to approach females; in this case the head is tucked in and the bill is tilted downward. A much more rapid surface movement that often produces a closer approach to females (in two of four cases) is one that Johnsgard (1966) called "motor-boating." In this display the male would suddenly begin "planing" over the water with his head erect, breast high, and tail flattened on the water; rapid alternate paddling movements generate a conspicuous bow wave and a trailing of waves in the water, similar to those made by a speeding motorboat. This also produces a noisy and rhythmical splashing sound and, like the display flight, may sometimes actually take the bird away from the group rather than toward it (Marchant and Higgins 1990).

The hunched rush ("aggressive chase" of Johnsgard 1966) is also exactly comparable to the corresponding agonistic display of North American ruddy ducks, with the neck and back feathers ruffled and the head held close against the breast as the bird threatens or chases a rival. Both sexes gape aggressively, and at close range a soft "tet-tet-tet" can be heard, which is uttered by females and possibly also by males.

The major sexually oriented displays of this species (outlined below) are, with one major exception (sousing), fairly clearly derived from comfort or prediving movements and often are difficult to separate from them.

Head pumping, a silent, vertical, and rather rapid pumping of the horizontally held head, occurs during situations of alarm. It typically occurs when the birds have been frightened and are swimming in an alert, tail-cocked posture and then serves as a diving-intention movement. It also occurs as an immediate preliminary to the sousing sequence. In this situation the pumping beings with the tail flat on the water. As the pumping progresses, the tail is gradually cocked up to beyond the vertical. When the tail reaches that position, the sousing display almost always begins. The average of 8 head-pumping

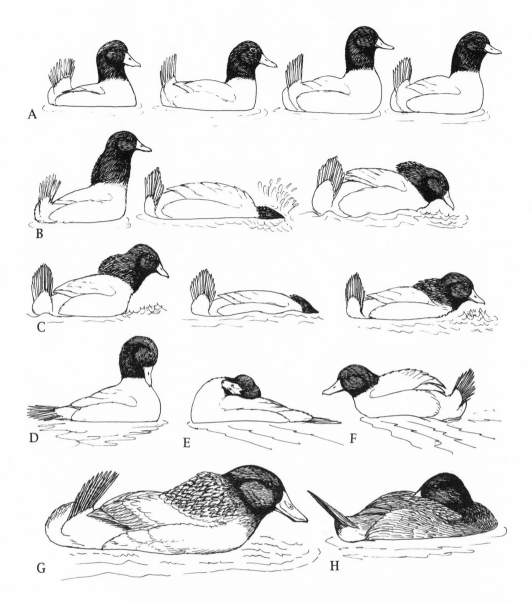

Figure 26. Social-sexual behavior of the male Australian blue-billed duck: (A) pumping phase of sousing; (B, C) choking phase of sousing; (D) dab-preening; (E) cheek rolling; (F) motor-boating (rushing); (G) detail of sousing; and (H) resting, or pseudosleeping. (After cine sequences and photos by Johnsgard.)

sequences observed by Johnsgard prior to sousing was 4.2, with a range of from 1 to 11.

Sousing (Fig. 26A–C) is performed periodically by males, and in Johnsgard's experience only in the presence of a female. However, in the maccoa duck and Argentine blue-billed duck it also is sometimes evoked by a sudden alarming stimulus, so it may also also occur in this species under similar circumstances. The sequence, which may sometimes last five seconds or more, begins with the male in a head-up, tail-cock posture, with the male usually facing the female. He then suddenly throws his head forward and downward, his neck inflated greatly, the associated neck feathers fully erected, and his tail still cocked (pumping phase). As the bill and throat strike the water he convulsively jerks his head back toward his breast (choking phase) and at the same time paddles with both feet to maintain a somewhat arched position of the body above the water surface. For several seconds the male continues a series of these jerky head movements, thus repeatedly splashing the water with his throat and bill. In 13 such sequences observed by Johnsgard, there were 3 to 7 jerks or chokes, averaging 5.1. The male then lowers his tail below the water and simultaneously lowers his head so that only the arched back is visible above water. In this posture, which may be held for a variable length of time of up to several seconds, the male may remain stationary or even may swim backward for a half meter or so. Finally, the head is raised and shaken, and the display sequence is terminated by a variable number (up to 5, average of 15 sequences was 2.5) of bill-dipping and head-shaking movements. Wing flapping has also been reported in conjunction with the choking phase of this display, but it was not seen by Johnsgard. However, the entire performance is accompanied by a very low frequency, throbbing call, consisting of a series of "dunk" notes repeated at regular intervals of 0.2 second (Marchant and Higgins 1990).

Dab-preening is one of the most common—perhaps it is the commonest—of the male courtship displays; it is the one that most often initiates a display sequence. From a normal swimming posture the male suddenly jerks his head up and down several times, simultaneously pressing his bill against the breast (Fig. 26D). The display is silent, but it is performed much more rapidly than is true of normal preening, and the displaying male tends to face the female. Although the dab-preens are usually given in short series of 3 to 6 separate preening movements, they sometimes may extend without interruption to as many as 20. Dab-preening is often followed by bill dipping, apparent exhaling (producing bubbles), and then flicking water laterally with the bill (bill dipping, head flicking), as is often also done during display by other *Oxyura* species.

Cheek rolling (or head rolling or rolling cheek on back) is performed frequently during display; it differs from normal comfort-movement activities only in its display context and the fact that a quite regular alternation of right and left movements occurs. Additionally, the orientation of the male is usually toward a female. Usually three to six such head movements are made, with the cheeks being rubbed rapidly over the scapulars of either side (Fig. 26E).

Lurching is a common display that sometimes is performed in conjunction with other displays, such as wing flapping and head rolling. The bird becomes momentarily rigid, back-paddles a short distance, and then suddenly lurches forward in the water as if it had been attacked from below. Usually there is little actual forward movement of the body, but a conspicuous gurgling and splashing sound is produced. This display is sometimes preceded by lowering the bill in the water, then suddenly and simultaneously blowing bubbles and kicking. It is usually followed by a bill dipping, head flicking sequence or by rapid wing flapping and may be one of the three most common of male displays (Marchant and Higgins 1990).

As an independent display, *wing flapping* is usually preceded by tail shaking and a single head-dipping movement. The wings are then quickly flapped a few (typically three) times, which produces an audible whirring sound. Wing flapping may also occur in connection with other displays, often serving as an apparent finale to a courtship sequence.

Dorsal preening (or preen dorsally) was not definitely observed by Johnsgard to be used in a display context, but it may also be ritualized. On one occasion it occurred between wing flapping and lurching; on another occasion it appeared to serve as an introduction to head pumping and subsequent sousing.

A combination of *bill-dip, head-shake/head-flick* movements often occurs during general display, seemingly as a preliminary action or as an end to other sequences. It may differ somewhat from comparable actions in other male *Oxyura* in that air is evidently expelled from the bill during the bill-dipping phase, causing bubbles to be produced (Marchant and Higgins 1990).

Two displays that were not observed by Johnsgard but were subsequently described by Marchant and Higgins (1990) were "whisking" and "look-each-way-and-dive." *Whisking* consists of a rapidly lateral tail shaking at the water's surface, creating a surface disturbance of bubbles and ripples and producing a soft thrashing sound. Apparently this display is very much like the corresponding tail-vibration display of the white-headed duck, although orientation toward a specific female is less apparent. According to Marchant and Higgins, it is one of the three most common of male displays, the other two being dab-preening and lurching. It is often followed by a rapid backwards kick with both feet simultaneously and a lowering of the bill to blow air into the water. Finally, the wings are flapped vibrantly.

The *look-each-way-and-dive* display begins with the male close to the female and facing her. His bill is held partially open and pointed downwards. The head is then rapidly switched from side to side just before the bird slips below the surface in a shallow plop-dive. He then emerges almost at the point of entry. A series of sharp "chee-chee-chee" notes is uttered during this display.

Other activities normally associated with comfort behavior,

such as wing-ruffling, swimming-shakes, and head-shaking movements, also sometimes occur during general social display (Marchant and Higgins 1990). Apparently coordinated diving behavior has also been reported by some observers (Johnsgard 1966). Some of these actions may well be ritualized and function as actual displays, as they certainly are in other *Oxyura* species, but current information is inadequate to determine these points. As with other stifftails, apparent sleeping ("pseudosleeping") may occur during social interactions (Fig. 26H), especially by females, but this book's authors do not believe that it represents a ritualized signal in any stifftails.

Fertilization Behavior

Few observations of copulation have been documented. Copulations have been observed by Shanks (1954) and Wheeler (1953). In Wheeler's account, the male chased the female underwater at high speed until he caught her, and copulation occurred with the female entirely submerged. Although this sounds atypical, recent observations by Rushton and Fullagar (summarized by Marchant and Higgins 1990) have apparently confirmed this as a typical mode. Males have been seen chasing females in a series of dives that ended with the male surfacing with an everted pseudopenis, suggestive but not positive proof that copulation has occurred.

NESTING AND PARENTAL BEHAVIOR

Nests and Nest Building

Nests are usually constructed of dead *Typha* leaves, typically in stands of mature cattails. They are either sited over water or sometimes placed on small islands. They are usually about 40 centimeters wide, with a domed roof about 40 centimeters above the base, a main entrance hole about 7.5 centimeters in diameter, and a thin spot in the vegetation for emergency exits. There may be a slight lining of down, but often there is none, and the incomplete clutch may be partially covered with vegetation when the female leaves the nest during the egg-laying period. The nest cup is deep and is built up from a platform of dead leaves or stems that is usually elevated 15 to 30 centimeters above water level. The nest is constructed entirely by the female, from materials gathered at the site. Sometimes the deserted nest of a coot or other waterbird may be used. Males may influence the site selection by the construction of several preliminary platforms. Nests made under conditions of captivity are sometimes placed as close as 2 meters apart; in the wild they have been found as close as 8 to 9 meters apart (Marchant and Higgins 1990), strongly suggesting nonmonogamous mating.

Egg Laying and Incubation

Eggs are laid at intervals averaging about 24 hours, but in larger clutches the later eggs may be laid at 48-hour intervals. Laying typically occurs early in the morning, usually shortly after dawn. Re-nesting following clutch loss occurs, but the renesting interval may vary from as short a period as 22 days to as long as 130 days. In captivity, 4 of 16 consecutive nestings were dump-nests, with from 8 to 13 eggs present, the females tending to lay all their eggs in a single host nest (Marchant and Higgins 1990). In nests that have not been influenced by dump-nesting, the commonest number of eggs is 5 or 6, but may be as few as 3 (Frith 1967). The incubation is entirely by the female and begins with the laying of the penultimate egg. At least in captivity, it consistently requires 24 days but has been estimated at 26 to 28 days with wild birds. Females typically leave their nest once a day, when they hurriedly forage for 10 to 15 minutes before returning again to incubate. When the eggs hatch, the eggshells are quickly removed from the nest and are apparently hidden (or possibly eaten), as they have never been located (Marchant and Higgins 1990).

Hatching and Brood-Related Behavior

The young typically hatch early in the morning, are brooded for about 24 hours on the nest, and then leave the nest early the following morning. For the first two days the female tends to hide the brood in reedbed fringes; she may return the ducklings to the nest for nighttime brooding. Platforms near the nest may also be used for brooding the ducklings. Although males may sometimes seemingly associate with the brood, this is done not to protect the brood, but rather to display sexually to the female or perhaps even the young. At least in captivity the female may brood her ducklings for about 12 days, but brood mergers may occur, and the ducklings increasingly forage for themselves as they grow older. They always feed themselves (unlike musk ducks), and their foraging dives average about 10 seconds. Intruders are threatened by the female, by hissing, gaping, lifting the scapulars, and foot splashing. The "portable" territory around the chicks is strongly defended for about 10 to 14 days, but such protection is progressively relaxed after 2 to 3 weeks. After 3 or 4 weeks, the young have become fairly independent of their mothers; they are fully independent by 60 days. Fledging may occur at about 60 days, and at least in captivity both sexes become sexually mature as yearlings. Captive females may have at least two broods, and up to a maximum of four, within a single 12-month period. The minimum intervals between the rearing of one brood and the onset of a second clutch have been determined as 55 and 56 days (Marchant and Higgins 1990). However, double-brooding in the wild has not been proven for any species of stifftail, although it has been suggested as possible for ruddy ducks.

REPRODUCTIVE SUCCESS AND STATUS

Little information exists on breeding success. At least in captivity females typically rear about 80 percent of their brood successfully, with losses usually caused by disease rather than predation. Wheeler (1953) observed that as many as five young may survive to eight weeks, but it has been estimated that on average only three ducklings per brood are likely to survive (Marchant and Higgins 1990).

The status of this species is difficult to determine, because of confusion with coots during winter aerial surveys and the species' preference for living in areas of dense vegetation during the breeding season. Flocks numbering in the several thousands have been seen in some favored wintering sites (such as Lake Kangaroo, Victoria). However, summer breeding-season surveys have generally turned up only a few thousand birds in hundreds of Victoria wetlands (a three-year average of 1847 birds on 477 wetlands), and blue-billed ducks composed less than 1 percent of all the ducks that were then counted (Marchant and Higgins 1990).

Male white-headed duck performing kick-flap display.

White-Headed Duck

OXYURA LEUCOCEPHALA
(Scopoli) 1769

Other Vernacular Names
> Spanish duck, spiny-tailed duck, stiff-tailed duck, white-headed stifftail, Ural duck; Weiskopfruderente (German); erismature à tête blanche, erismature leucocephale (French); gobbo rugginosoa (Indian); malvasia, pato de cabeza blanca, pato porrone pato, tarro (Spanish); savka sinonossaia (Russian); kaczka bialoglova (Polish).

Range of Species (Fig. 27)
> Local resident in the Mediterranean region, but now highly threatened and limited to southern Spain (Andalucia). The species is also a rare resident along the northwestern coast of Africa (Algeria and perhaps Morocco, occurring during winter in Tunisia). Also breeds or has bred historically from Turkey east through southern Russia and the central Asian steppes, probably rarely reaching as far as Dzungaria in China (Junggar Basin), western Mongolia, and Tuvinskaya in Siberia. The eastern breeding population now is evidently largely confined to southern Russia and Kazakhstan and winters almost entirely in central Turkey (mainly on Lake Burdur Gölu). Local wintering also occurs in Israel, Iran, and Pakistan. Wintering once occurred in the Nile Valley (no recent records), Iraq (few recent

Figure 27. Distribution of the white-headed duck: The species' breeding range is shown hatched, the areas of denser population are inked in, and the wintering range is shaded.

sightings), and, rarely, northern India. Wintering birds may still occur in small numbers in Romania and along the Caspian Sea. Conservation and reintroduction efforts have been made in Spain, Italy, France, and Hungary.

Measurements (Millimeters)

Wing: males 157–172 (average of 10, 162), females 148–167 (average of 6, 159). **Culmen:** males 43–48 (average of 17, 45.5), females 43–46 (average of 16, 44.5). **Tail:** males 85–100 (average of 9, 92.4), females 75–93 (average of 5, 85.5). **Tarsus:** males 35–38 (average of 16, 35.9), females 33–37 (average of 16, 34.9) (Cramp and Simmons 1977). **Wing:** 8 males 164–172, 14 females 150–167. **Culmen (exposed):** 8 males 46–47, 14 females 43–46. **Tarsus:** 8 males 43-46, 14 females 41-45 (Ali and Ripley 1968).

Measurements of birds of the eastern and western populations indicated significant differences only in bill length and bill depth at base, which were suggested by Amat and Sanchez (1982) as possibly being related to the competitively less harsh environment faced by western birds (estimated at five total interacting species versus nine species in eastern populations).

Weights (Grams)

Wild adults from the (former) USSR, males in April ranged from 720 to 800 (average of unstated sample, 755) and in June-July from 600 to 760 (average of 3, 703). Three females in April ranged from 720 to 900 (average 836.6), a July female was 620, and females in September ranged from 510 to 820 (Dementiev and Gladkov 1952; Cramp and Simmons 1977). Three Pakistan males in December ranged from 553 to 865 (average 737.3), three females 539 to 631 (average 593) (Savage 1965). Captives in England, three males 635 to 770 (average 715.0), nine females 525 to 710 (average 625.6) (Carbonell 1983). These inadequate samples suggest a mean male weight of about 717, a mean female weight of about 657, and a mean male-to-female mass ratio of about 1.1:1. Using the square of the mean wing lengths as an alternate mass estimate, this predicted ratio is 1.04:1.

DESCRIPTION

Natal Down

The down is dark brown, almost black, on the head and upperparts, with a white line below each eye and on the chin and sides of the neck. There are no dorsal markings but feathers are paler on the rear edges of the wings, and the underparts are whitish. Tail feathers are dark brown and stiff. The iris is brown; the bill (already thickened basally at hatching), feet, and legs are black.

Juvenal Plumage

This plumage is much like that of adult females, but there is no chestnut tinge to the feathers, except for some on the barred upper-tail coverts; mantle and scapulars are dark grayish-brown, finely vermiculated and speckled with pink-cinnamon. The head and hind-neck are browner than in adult females, the white line below the eye is broader and paler. The chin, upper throat, and sides of the upper neck are pale gray; the lower neck is mottled with sooty. Undertail coverts are gray to whitish; the shafts of tail feathers are bare near the tip and often have downy tips. The sexes are similar, but males have a rusty cinnamon cast on the chest, and females have a more uniform pattern on head and neck. The iris is pale gray, light brown, or yellow-brown; the bill is light leaden gray; the legs and feet are pale cinnamon (Cramp and Simmons 1977).

First-Winter Plumage

This has not yet been described in detail but is generally femalelike, immature males being redder on the back (Ali and Ripley 1968). Through the first winter juvenile birds may be identified by their narrower and more spike-tipped tail feathers. By their first spring, young males increasingly resemble adults, with their body plumage and bill color approaching the adult condition, but they have mottled dusky heads (Madge and Burn 1988).

First-Year Breeding Plumage, Male

First-year males are highly variable in appearance as to head patterning (Amat and Sanchez 1982; Carbonell 1983), as illustrated in Fig. 18. Judging from these observations, there are not only first-year (subadult) males with entirely black heads, there are some with adultlike head plumages. Additionally, some subadult males at the Wildfowl Trust acquired white feathers on their cheeks as the breeding season progressed, whereas others retained a large amount of black on their cheeks throughout their second year. Bill color of subadult males changes from bright blue during the breeding season to dark brown in the nonbreeding season. Contrary to some earlier descriptions, immature males also normally have brown eyes (Carbonell 1983), although some subadults may have a yellow iris color (Amat and Sanchez 1982).

Definitive Breeding and Nonbreeding Plumages, Male

No definite seasonal plumage changes are present, but individual variation occurs in the upperparts (pale reddish-gray to dark reddish-brown) and in the amount of white present on the face. Additionally, birds from eastern populations tend to have dark reddish-brown, rather than yellowish-gray, sides and underparts as is reportedly typical of western populations. The crown is black, the rest of the head white. The neck is black, becoming rusty-gray or chestnut on the mantle, the feathers marked with fine black lines. The back and rump are gray, with indistinct black barring, merging with more chestnut-colored uppertail coverts. The lower neck, breast, sides, and flanks are chestnut to yellowish-gray, the color becoming paler below and posteriorly, and merging with the silvery-white abdomen and undertail coverts. The tail is black; the wing gray dorsally, with vermiculations on the coverts; underwing coverts are mostly gray, with white axillars. The iris is brown, not yellow or grayish-white as sometimes described (e.g., Dementiev and Gladkov 1952). The bill is bright blue during breeding, the color fading to dark brown in the nonbreeding season. Legs and feet are ashy-brown to lead-gray.

Definitive Breeding and Nonbreeding Plumages, Female

No evident seasonal plumage changes are present, although there is great individual variability in the cheek markings, which can range from a straight white line below the eye to an almost completely dark-brown cheek with only spots of white feathers. The female differs from the male mainly in lacking clear white on the head and in being generally more rufous, especially on the underparts (at least in the western European populations). Crown, nape, and cheeks are brownish-chestnut, with blackish barring. A white streak usually extends from below the eye almost to the nape; the rest of the head and foreneck are dotted or streaked with blackish. Upperparts are generally chestnut-toned but sometimes more grayish. Otherwise, the female is like the male. Old females may develop a variable amount of white on the head, up to about 60 percent in some birds (Hughes, pers. com.). The iris is dark brown; the bill dull leaden; the legs and feet grayish-black (Phillips 1922–26).

IDENTIFICATION (FIG. 32C)

In the Hand

The bird is easily distinguished at all ages and in any plumage from all other *Oxyura* by its very long tail (75 to 100 millimeters) and also by its highly swollen bill, which in adults is at least 43 millimeters long.

In the Field

Over nearly all of its native range this is the only species of native stifftail and thus should be easily recognized by its long tail and distinctively swollen bill. However, now that North American ruddy ducks have reached southern Spain (where contact has occurred and hybridization has recently been reported), possibilities of confusion exist. Adult males differ from those of North American ruddy duck in that, besides their much larger bill, they usually have the white facial patch extending to the forehead and above the eye, whereas in ruddy ducks the white cheeks barely reach the lower edge of the eye. Male white-headed ducks are generally relatively pale grayish-brown on the upperparts and more silvery on the sides, rather than uniformly deep chestnut as in ruddy ducks. Females have a similar pale plumage cast, with darker fine penciling rather than distinct brownish barring on the sides and back.

ECOLOGY

Habitats and Densities

During the breeding season, the favorite habitats of this species consist of brackish lakes and lagoons that are fringed with emergent herbaceous vegetation, such as reeds (*Phragmites*) and cattails (*Typha*), and have a good growth of pondweeds (Potamogetonaceae). However, during winter it is found in large, alkaline, shallow lakes, or sometimes even at sea, often in rocky bays but not in sites subjected to heavy wave action. Lakes are chosen that are sufficiently alkaline that they do not normally freeze over (2000 to 8000 parts per million of dissolved salts) and have little emergent vegetation but are still able to support algae and submerged aquatics such as wigeongrass *(Ruppia)* and pondweeds. In Spain the birds concentrate in areas of the lake where the water is less than a meter deep, but in Turkey the birds may forage to a depth of about 15 meters (Hughes, pers. com.). In the essentially nonmigratory Spanish population, lagoon depth was the single measured variable that significantly separated lagoons that were used by these birds from those not used. Additionally, general water alkalinity (carbonate) levels tend to be positively associated with the presence of the species, whereas increased phosphate levels in the water tended to show an inverse correlation with this species' use (Amat and Sanchez 1982). In the former Soviet Union the species' populations increased sharply on steppe lakes during periods of drought, when saline levels also increased, but declined during years of high water levels (Dementiev and Gladkov 1952).

Breeding densities have evidently been extremely low everywhere in recent years, probably as a reflection of the species' generally low overall population levels. Wintering populations in Andalusia varied greatly between 1972 and 1981 (Amat and Sanchez 1982), but no useful estimates of breeding densities are yet available for that region. In Kazakhstan breeding-density estimates of from 0.3 pair per 100 hectares of breeding habitat in the north to as many 25 pairs per 100 hectares in the south have been made (Cramp and Simmons 1977). However, populations in Kazakhstan have reportedly declined seriously in recent years (Collar and Andrew 1988).

Foods and Foraging
Very few data exist on actual foods consumed. Birds collected from the former Soviet Union had eaten the leaves of pondweeds, eelgrass *(Zostera)*, and stonewort algae *(Chara)*, plus the seeds of pondweeds, wigeongrass *(Ruppia)*, bulrushes *(Scirpus)*, and a few other other aquatic or semiaquatic plants. Insects that are consumed include the larvae of midges (Chironomidae), as well as some mollusks and their eggs *(Hydrobia)*, in addition to crustaceans (Dementiev and Gladkov 1952). Elsewhere in their range insects such as waterboatmen (Corixidae) have been reportedly eaten by females, and ducklings evidently eat insects almost exclusively (Cramp and Simmons 1977). More recently, the food contents of two adult specimens were found to contain mainly chironomid larvae, as well as some seeds and softparts of various aquatic plants, odonate nymphs, and some gastropods and cladocerans (Torres and Arenas 1985).

Foraging is done by repeated diving; some foraging dive durations obtained at the Wildfowl Trust were presented in Table 6. Additionally, Matthews and Evans (1974) reported that the birds foraging in Wildfowl Trust pools no more than 2 meters deep remained submerged for as long as 40 seconds. Mean dive duration for 200 dives was 21.9 seconds, and mean surface pauses (600 timed) lasted 7.6 seconds. Up to 50 such foraging dives occurred in rapid succession, and these were separated by intervening surface pauses of 5 to 8 seconds. Amat and Sanchez (1982) reported on foraging dives by wild ducklings and adults in a variety of Anadalusian lagoons. Among ducklings of varied ages the dives ranged in average lengths from 16.2 to 32.4 seconds; the average intervening pauses ranged from 7.3 to 12.2 seconds. Among adult males, average dive durations ranged from 24.6 to 43.6 seconds, with intervening pause intervals of from 9.2 to 15.4 seconds. Among females, dives averaged from 21.7 to 43.5 seconds, with mean pause intervals of from 7.7 to 14.9 seconds. The observed interlocational differences were attributed by the authors to different water depths in different lagoons. The intersexual differences in diving durations were statistically significant for most locations, although average pause-interval differences were not.

Competitors, Predators, and Symbionts

Amat and Sanchez (1982) believed that interspecific competition in this species might be strong enough to explain somewhat differing bill measurements in the species' two major population components, through ecological character displacement. However, there is no comparable case of such intraspecific differences in any other waterfowl, and the evidence for this hypothesis seems relatively slight at best. Other explanations, such as differences in average salinity levels encountered in the two populations, should probably also be considered when evaluating interspecific and intraspecific bill-shape differences of stifftails generally. Amat (1984) tried to evaluate possible competitive interactions between white-headed ducks and two other sympatric diving ducks in Andalusia, the common pochard (*Aythya ferina*) and the tufted duck (*A. fuligula*). The results of these studies were summarized earlier (Chap. 3) and need not be repeated here. Suffice it to say that considerable overlap occurred among the three species in terms of all three parameters measured—namely, the relative distances from shoreline that were used for foraging, the mean underwater distances covered between diving and emergence during foraging dives, and the general foraging strategies used (surface-foraging or near-surface-foraging versus diving). Of greater concern than any of these species as competitors with white-headed ducks is the recent appearance of North American ruddy ducks in Spain (see Reproductive Success and Status below). This unwelcome development may prove to be a major threat to the continued existence of the white-headed duck there.

No specific information on general predators exists, although in Spain eggs may reportedly be taken by water rats (*Arvicola sapidus*) and by marsh harriers (*Circus aeruginosus*).

ANNUAL CYCLE

Movements and Migrations

Movements in the European (Spanish) population are quite limited. After the reproductive season, most of the Andalusian birds congregate at a single lagoon (Laguna de Zoñar, Cordoba Province), where they form monospecific groups that reach an annual peak average of about 10 individuals during November (Amat and Sanchez 1982). The numbers drop off gradually to a low of less than two individuals during May, presumably diminishing as the birds disperse to breeding areas, all of which are located within about 200 kilometers of Cordoba. It is possible that some interchange also occurs between the Spanish and northwestern African populations, especially during years of drought in Spain when major migrations might occur (Hughes, pers. com.).

Much longer migrations are typical of the more easterly, Central Asian breeding population, much of which is probably now largely limited to Kazakhstan. These birds begin to leave their breeding areas by late September and early October, with concentrations on the eastern side of the Caspian Sea and Turkey correspondingly increasing during late October and November and in Pakistan during November. The return migration begins in February, with birds having departed Pakistan and Iraq by late April and the major spring arrival in Kazakhstan occurring at the end of April and the beginning of May. Even those birds breeding in western Siberia have reached their breeding areas by mid-May (Dementiev and Gladkov 1952; Cramp and Simmons 1977).

Molts and Plumages

Some information on molting is presented in Description above. Because of the absence of obvious seasonal adult plumage changes, little information on molting is available on wild birds. However, in southern Spain the males all lack blue bills during October and November. From December until March there is a rapid increase in the proportion of blue-billed males, and all of the males exhibit blue bills from March until August (Amat and Sanchez 1982).

At the Wildfowl Trust, white-headed ducks underwent two flightless periods yearly, during which they also replaced their rectrices. The prebreeding molt began in January, with the bill color of males changing first. The primaries were then shed, followed about a week later by the rectrices. Dominant males regained the blue color of their bills and molted their feathers before nondominant ones, which started to lose their feathers at least a month later. By March, all the males had blue bills. The postbreeding molt began in July, with the primaries being lost first. While flightless, the birds lost their rectrices and finally their bill colors faded. The nondominant males molted and lost their blue bill colors about a month ahead of the dominant males. At least one female that had raised her own brood lost her rectrices before becoming flightless. The primaries required two to three weeks to regrow, and the rectrices three to four weeks. In two cases a female lost all but her two outermost rectrices during the prebreeding molt and did not grow any new ones until she finished breeding.

Breeding Cycle

Relatively few data are available for wild birds, but in Spain, North Africa, and the former Soviet Union eggs are generally laid during late May and June (Cramp and Simmons 1977; Dementiev and Gladkov, 1952). In Andalusian Spain, 24 laying dates extend from the early May to mid-July, with a peak during the period May 10 to May 30, and a possible subsidiary peak (perhaps reflecting renesting efforts) in mid-June (Amat and Sanchez 1982). Egg laying by captive females at the Wildfowl Trust usually extended from mid-April to mid-July during three years of study. This egg-laying period was longer than that of the North American ruddy duck but shorter than those of the maccoa duck and Argentine blue-billed duck. The laying of second clutches was regular in these females, whose eggs were removed for separate incubation a few weeks after incubation had begun. Under these conditions some females laid a total of three or perhaps even four clutches within a period of four months. This large energy drain associated with persistent renesting may have accounted for the relatively early termination of breeding in females of both these Northern Hemisphere species (Carbonell 1983).

SOCIAL AND SEXUAL BEHAVIOR

Mating System

The typical breeding system of white-headed ducks remains in some dispute. Observations of captives at the Wildfowl Trust and elsewhere (Matthews and Evans 1974; Veselovksy, 1976; Carbonell 1983) strongly suggest that males are polygynous and form no pair bonds. However, Amat and Sanchez (1982) believed that males do form pair bonds through their defense of a nesting territory and, thus, of a specific female. In view of the very high proportion of daily time (about 14 to 15 percent) that males at the Wildfowl Trust spend in courtship display (see Table 3), which was the highest of any stifftail studied by Carbonell, and even slightly higher than the display intensity of the distinctly polygynous maccoa duck, it seems likely that pair bonds are probably not normally present. In one case observed by Carbonell, a single male dominated the others on the pond, which were chased not only by him but also by the females. These females would not tolerate being courted by the nondominant males. In another pond, only one male had the attention of the four females during two years. By the third year of study a second male had become accepted by two of the females, while the first male was then associated with three. However, males and females of pre-sumed pairs did not remain together except when the male was displaying or looking for suitable nest sites. As soon as the female began to incubate, the fragile association was dissolved.

Hughes (pers. com.) has observed that the dominant male may change during a single breeding season. In a pen containing two males and two females of the white-headed duck and two females of the North American ruddy duck the dominant male changed two months into the breeding season. As this occurred, the males also switched plumage types, with the subordinate males becoming brighter in plumage and bill color and the previously dominant male becoming duller. As he gained in color, the male was allowed more access to the females until he became dominant.

Territoriality

Matthews and Evans (1974) did not describe any territoriality in the birds they observed at the Wildfowl Trust. Furthermore, observations by Carbonell (1983) and Johnsgard (unpublished) have not indicated evidence of male territoriality in this species. However, as mentioned above, Amat and Sanchez (1982) suggested that males defend breeding territories within which their females are present, since they observed males swimming near active nests in a patrolling manner. No estimates were provided by these authors as to territorial sizes or any other physical characteristics.

Courtship and Pair-Bonding Behavior

The first detailed description of displays in this species was by Matthews and Evans (1974); earlier accounts (e.g., Dementiev and Gladkov 1952) are generally unreliable and have prompted unjustified comparisons with the North American ruddy duck. Brief accounts have also been provided by Veselovsky (1976) from captive and wild birds of the eastern population. Amat and Sanchez (1982) and Torres et al. (1985) have provided good accounts, the latter accompanied by fine drawings, based on their observations of wild birds in Spain.

The average duration of male display in this species was 4 minutes in Carbonell's observations; the maximum time that a single male was observed to display without interruption was 20 minutes. The forms and durations of the displays described below are based on a maximum of four males.

The ringing rush was first reported by Amat and Sanchez (1982) and was later noted at the Wildfowl Trust by Carbonell (1983), although it was apparently never seen by Matthews and Evans (1974), Veselovsky (1976), or Torres et al. (1985). Neither Johnsgard nor Hughes (pers. com.) has observed it, and Hughes believes that its occurrence in this species is questionable.

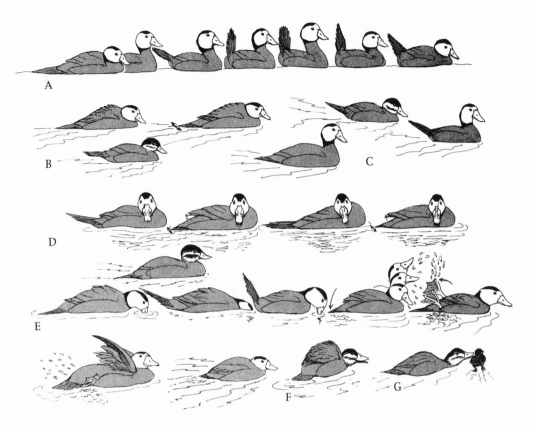

Figure 28. Social-sexual behavior of the male white-headed duck: (A) head-high, tail-cock posture, (B) sideways hunch with tail vibration, (C) flotilla swimming, (D) sideways piping, and (E) kick-flap sequence. Also shown is female (F) in open-bill threat and (G) touching duckling's head with bill. [After sketches from cine sequences by Carbonell 1983 and photos by Matthews and Evans (F, G) 1974.]

The *head-high, tail-cock* (Fig. 28A) display represented 36.7 percent of the total display activities tallied. It lasted for from 1.7 to 2.5 seconds (average of 6, 1.96 seconds). In this display the male raises his head and closes his tail simultaneously and rather slowly and then returns his head and tail to the normal position. Of 55 complete courtship sequences recorded, 30.9 percent were initiated, and 21.8 percent terminated, with this display. The male may perform the display several times in an uninterrupted sequence, to a maximum of 15 observed; it is performed laterally to while directly in front of the female. This same posture is used during alarm and is also part of flotilla swimming.

The *sideways-hunch* display (Fig. 28B) represented 36.7 percent of the total displays tallied; it is the primary display of the white-headed duck. The male positions himself laterally while in front of the female. With tail spread, lying flat in the water and vibrating, the male holds his head in a low swimming position; his scapulars are slightly raised. In this posture the male utters a mechanical, "tickering purr" and paddles his feet quickly. If more than one female is nearby he may switch from one female to another while remaining in the same posture and without stopping the vocalizations. Occasionally, if the female seems "interested," the male may turn in a half-circle to exhibit his other side. Of 55 full courtship sequences recorded, 67.7 began with, and 21.8 percent finished with, the sideways hunch.

Together with sideways piping (described below), the *kick-flap* display (Fig. 28E) represented 11.5 percent of the total display activities tallied. It lasted from 0.6 to 0.9 second (average of 5, 0.71 second). In this action the male sinks his bill and most of his head in the water, as well as his tail, while lifting his scapulars. Then, the tail is lifted, the bill is flicked in the air, the tail strikes the water, and both feet are kicked, producing a splash of water. The male then immediately assumes the sideways-piping posture. Typically, kicking occurs only once in the display, although Matthews and Evans (1974) reported it occurring twice. Double-kicks may sometimes occur in immature birds, whose coordination is still poor.

The *sideways-piping* display (Fig. 28D) immediately follows the kick-flap, with the male positioned laterally to the female. It lasts 0.7 to 0.9 second (average of 19, 0.83 second). The head and bill are directed toward the female, and the tail is spread and slightly lifted initially. Then the tail is lowered and vibrated on the water surface; at the same time the folded wings are raised to the center of the back, while the bill is opened and a piping sound is uttered. Of 24 such sequences by adult birds, the average number of piping calls was 13.4 (range 7 to 23). A second-year male produced only 1 or 2 such pipings. The average duration and limits noted above correspond to these individual piping elements. The kick-flap, sideways-piping sequence was always performed immediately after the sideways hunch, except for a few occasions when a head shake was inserted between them.

The *head-dipping* (or dipping, or dip-diving) action is clearly derived from bathing, as was described earlier. Several previous authors considered it to be ritualized, but observations by Carbonell do not support this view. In all of her observations, head dipping led to preening rather than to other social displays.

Head shaking (or "sideways flicks" of Matthews and Evans 1974) represented 11.5 percent of the actions tallied during courtship. The action is identical to that used as a comfort movement. However, 42.4 percent of the head shakes preceded, and 54.5 percent followed, the sideways hunch, suggesting that it has become part of the species' display repertoire.

Other authors have reported various other comfort movements as part of the display repertoire as well, including dab-preening, wing shuffling, and cheek rolling, but during Carbonell's observations the movements were not clearly linked to display activities. Simultaneous diving by male and female have also been observed in wild birds (Amat and Sanchez 1982), but such diving as was observed at the Wildfowl Trust was not clearly done in a social context.

Synchronized swimming—*flotilla swimming* (Fig. 28C)—by males and females was observed several times by Matthews and Evans (1974) but was seen only occasionally by Carbonell. The birds would swim up and down the pond, often changing relative positions, while in the head-high posture, their tails held flat on the water or only slightly cocked. The social significance of such swimming in this species remains conjectural.

Fertilization Behavior

Three copulation sequences have been observed by Carbonell (1983); Torres et al. (1985) have also described this behavior. In all three cases observed by Carbonell the male was initially performing the sideways-hunch display with the female swimming next to him. The male then performed the kick-flap, sideways-piping display, and with each successive piping series came closer to the female. She did not move away, and after a series of perhaps 5 to 10 pipings, she became prone and the male mounted. Copulation lasted three to six seconds, and after treading had been completed both sexes preened. On one occasion the male then displayed to other nearby females. No specific precopulatory behavior was observed in the female before becoming prone. In the account of Torres et al., sideways piping by the male and neck stretching by the female preceded the copulation.

NESTING AND PARENTAL BEHAVIOR

Nests and Nest Building

Nests of this species are typically built of emergent vegetation and consist of a cupped platform of stems and leaves, over which a roof may be formed by bending down overhead leaves. Old nests of coots or of ducks such as tufted ducks may also be used. Nests at the Wildfowl Trust have been placed in a variety of sites. Of 72 nests or attempted nests, 61.1 percent were placed among rushes, either on artificial rafts made of bamboo and reeds or in clumps of rushes along the water's edge. Additionally, 25 percent of the nests were in nest boxes, and 13.9 percent were in other sites, mostly under rose bushes, but one was in a clump of dock *(Rumex)*. The nests placed inside boxes were constructed of small twigs, and those among rushes had partial overhead roofs of bent-over stems. One nest had some down, but most had no down at all. Dimensions of white-headed nests found at the Wildfowl Trust are presented in Table 12. Of four nests found in Andalusia by Amat and Sanchez (1982), all were situated on old coot nests, within dense cattail stands. Measurements given for these wild nests averaged somewhat smaller but were not otherwise greatly different from those presented in Table 12.

Egg Laying and Incubation

Little information exists for wild birds. Amat and Sanchez (1982) reported that four Spanish nests had 5, 6, 8, and 10 eggs present, representing an average clutch of 7.3 eggs. Eggs are laid at approximately 1.5-day intervals, and as many as 15 have been found in a single nest, suggesting that dump-nesting sometimes occurs (Cramp and Simmons 1977). At the Wildfowl Trust, Carbonell (1983) estimated that females spend about 87.3 percent of their daytime foraging during the first day of laying and on the day prior to it. While incubating, females spent 92.7 percent of the daytime on the nest, with incubation bouts averaging 4 hours (3 observations) and periods off the nest averaging 16 minutes (22 observations). The mean clutch size at the Wildfowl Trust during Carbonell's studies consisted of 5.7 eggs (Table 15); 5 or 6 eggs were the most frequent clutch sizes. Clutch size was positively correlated with the age of the female. One case of interspecific parasitism was observed, when an egg was placed in the nest of an Indian spot-billed duck (*Anas poecilorhyncha*) who was incubating 9 eggs of her own. This particular egg was laid by a yearling female and may have been the first that she ever laid. The average interval between successive clutches laid by the same female was 17 days, and as noted earlier, some females laid at least three clutches, and perhaps four, within a single nesting period. The incubation period was determined by Carbonell to be 24 to 26 days, which is in agreement with an earlier estimate of 25 days by Matthews and Evans (1974). Newly hatched ducklings averaged 56 grams (Table 14).

Hatching and Brood-Related Behavior

The best information on brood-related behavior comes from Carbonell's (1983) observations of a female that produced two broods of five ducklings during two successive years. The female did not remove the eggshells from the nest, but rather they remained there and were crushed by the adult bird. It took 9 minutes for one female to get the first three ducklings of her brood out of the nesting box and into the water for the first time. They spent 12 minutes swimming about before the female returned to the nest to collect the others. Only one of them emerged; the fifth and last remained in the nest in spite of the female's calls and visits. After only 7 minutes on the water the young ducklings began to dive, while the female remained nearby, swimming and constantly watching for danger. The ducklings uttered a soft "cheep-cheep" while following their mother, and the female in turn constantly uttered a "gek-gek" call. She would stretch her neck forward above the water to elicit a grouping response to the chicks; with any disturbance she would call louder and faster while in a head-high posture. With a sudden or intense alarm she assumed the head-high, tail-cock posture, and the chicks would quickly dive—or would sometimes assume a similar head-high, tail-cock posture. Females chased all other birds that approached their broods including other white-headed ducks. Chasing was done by assuming the open-bill-threat posture (Fig. 28F). If the disturbance persisted, she performed the hunched rush, and at times actually attacked and fought the intruder. Other white-headed ducks exhibited great apparent curisity over small ducklings, and several adults would swim toward newly hatched young, despite the constant threats of the mother. However, none of the males performed courtship displays toward the chicks, as was observed by Matthews and Evans (1974). Except for one second-year male, which was observed trying to peck the swimming ducklings, none of the adults threatened or attacked the chicks.

At times females would approach a duckling only a few days old and, with bill slightly opened and neck outstretched, touch it lightly and briefly on the head (Fig. 28G), which presumably helps to strengthen family bonds. Missing or inattentive ducklings may be called by the female with a low, rattling purr, and a warning call "huu-rugru-u" may be uttered before diving to escape an avian predator (Mathews and Evans 1974; Cramp and Simmons 1977). Carbonell observed that very young ducklings were brooded under the wing while the female was on land or resting on a raft. Her vigilance and association with the chicks faded gradually as they grew older. By the eleventh day after hatching, the chicks were spending some time alone; when about 20 days old they were fully independent. Matthews and Evans (1974) reported independence by 5 weeks, and Amat and Sanchez (1982) by 15 to 20 days, so there must be considerable variation in this trait. Merged broods of ducklings were also observed among wild broods of white-headed ducks by Amat and Sanchez.

During their first week of posthatching life, the ducklings spent a surprising amount of time in the nest, but this was probably an artifact of data-gathering, perhaps a result of the relatively large number of hours spent observing them during the first day after hatching. The first night after hatching the young were led back to the nest, but from the second day onwards they did not return. Feeding by ducklings was done mainly by diving, although some surface-straining of foods was also performed. Average dive durations for white-headed ducklings were summarized in Table 7. These durations average somewhat shorter than do comparable figures for maccoa ducks and North American ruddy ducks. They are also somewhat shorter than mean dive durations reported for ducklings by Matthews and Evans (1974) for the first seven weeks following hatching. These authors observed some limited use of the nest as a resting site until the twenty-third day after hatching. By the second and third weeks the majority of the ducklings' time was spent in foraging (Table 5). During the first week the female spent 22.7 percent of her time foraging and 16.2 percent in watching over the brood (Table 4). After their chicks had become fully independent, females spent more time resting and correspondingly less time feeding. Male ducklings were not seen displaying toward adult females, although Matthews and Evans observed a male duckling displaying toward his own mother on several occasions when he was between 32 and 42 days old.

REPRODUCTIVE SUCCESS AND STATUS

Little information is available on reproductive success. Torres and Raya (1983) reported that, of 54 hatched ducklings, only 35.2 percent survived their first two months after hatching, with the amount of water contamination apparently influencing local survival rates.

This species is regarded as endangered in Spain. It is probably threatened throughout the rest of its Eurasian range and has been listed as "vulnerable" by the International Council for Bird Preservation (King 1990). Winter counts during the 1980s suggested that the world population is probably 10,000 to 15,000 birds, of which about 10,000 winter in Turkey; 1000 in Tunisia; 1000 in Pakistan, 800 in the Caspian Sea area; and 100 in Romania (Madge and Burn 1988). In 1993 the total world population was estimated at 19,000 birds, with 78 percent of the breeding birds believed to be in the former Soviet Union, 10 percent in Turkey, 4 percent each in Spain and Iran, 2 percent in Algeria, and 2 percent elsewhere (Green and Anstey 1992).

In Spain, where breeding is confined to Andalusia, the population was judged to be about 20 to 30 individuals in the late 1970s. This resident Spanish population has been increasing through intensive conservation efforts (protection of the Analusian wetlands) and reintroduction efforts using captive breeding stock, with about 140 birds released between 1988 and 1992. A maximum count of 786 birds was obtained in Spain during the winter of 1992 (de Vida 1993). Reintroduction has also been attempted in Hungary and has been proposed in Italy.

The wintering population of white-headed ducks is strongly localized, with an estimated 62 percent in Turkey, 22 percent in the former Soviet Union, 4 percent in Spain, 3 percent each in Israel and Iran, 2 percent in Pakistan, 1 percent each in Algeria and Tunisia, and 2 percent elsewhere (Green and Anstey 1992). Virtually the entire Turkish winter population (a maximum of about 11,000 birds in 1991) is confined to a single lake, Burdur Gölü; of these perhaps as many as 150 pairs remain to breed in Turkey. Burdur Gölü is an alkaline and saline closed-basin lake very rich in chironomid larvae and consequently an excellent wintering habitat. Recently, sport hunting has been banned on Burdur Gölü as well as on Yarisli Gölü, a secondary wintering lake. This is an important step forward, since hunting from speedboats had been a popular sport that was jeopardizing the white-headed duck. However, in the winter of 1992 to 1993 a maximum of only 3010 birds was reported on the two lakes and as many as 2000 may have been shot that winter in illegal hunting. Other apparently important wintering areas are Lake Occhali, Punjab Province, Pakistan, where 667 birds were observed in 1987, and Azerbaydzhan, where 3520 birds were seen during a winter survey in 1991.

One recently developing threat to the white-headed duck in Europe is the expansion of the range of the North American ruddy duck. This species, which first escaped from captivity at the Wildfowl Trust in 1952, was initially reported from continental Europe in 1965. Since then it has reached Spain, and from December 1989 to July 1991 there were 29 records of North American ruddy ducks from 10 wetlands in Andalusian Spain. Additionally, captive-bred hybrids between these two species were first documented at the Wildfowl Trust, but more recently at least 20 wild hybrids have been observed in Spain. It is known that such hybrids are fertile, at least when they are bred back to white-headed ducks. Additionally, both male ruddy ducks and male hybrids are socially dominant over male white-headed ducks during courtship, so it is possible that the North American ruddy duck may pose the most serious threat to the continued survival of white-headed ducks in Spain (Anon 1993; Sylvestre 1993).

Male musk duck performing whistle-kick display for female.

Musk Duck

BIZIURA LOBATA
(Shaw) 1796

Other Vernacular Names
> Diver, lobed duck, mould goose, steamer duck; Scharbenente, Kehllappenente, Lappenente (German); canard à membrane (French); pato almizclero de Australia, malvasia de papada (Spanish).

Range of Species (Fig. 29)
> Resident in southwestern Australia from Shark Bay to Cape Pasley, and from eastern South Australia (Spencer Gulf and Lake Eyre basin) east throughout Victoria and New South Wales to southeastern Queensland (Hervey Bay). Also resident in Tasmania. No subspecies recognized currently; the eastern population has at times been separated as *menziesi*.

Measurements (Millimeters)
> **Wing:** males 205–240 (average of 219, 223), females 165–202 (average of 248, 185). **Culmen:** males 36–47 (average of 240, 41), females 31–41 (average of 289, 35) (Frith, 1967). **Tail:** males 106–120 (average of 4, 113.5), females 99–107 (average of 4, 103.2). **Tarsus:** males 48.4–55.9 (average of 7, 52.6), fe-males 42–49.3 (average of 5, 44.8) (Marchant and Higgins 1990).

Weights (Grams)

Figure 29. Distribution of the musk duck: The species' breeding or residential range is shown hatched; the areas of denser populations are shown cross-hatched.

Males 1811–3120 (average of 243, 2398), females 993–1844 (average of 292, 1551) (Frith 1967). These figures represent a mean male-to-female mass ratio of 1.54:1. If the square of the mean wing length is used as an indirect mass estimate, the mean predicted mass ratio is 1.45:1. Females during pre-breeding period, average of 144, 1338; during prelaying period, average of 21, 1665; during laying period, average of 13, 1641; during incubation, average of 4, 1397; during brood-rearing, average of 21, 1223; during nonbreeding season, average of 98, 1346.

DESCRIPTION

Natal Down

The head, neck, and upperparts are brownish-black, grading to dull whitish-gray below, with no facial or dorsal markings except for a paler area around the base of the bill. The iris is blackish-brown; the bill dark gray to black, the upper mandible more grayish basally and the lower mandible light orange to orange; the legs and feet are black.

Juvenal Plumage

This plumage is similar to an adult's but with wider white to cream feather edges on underparts. Tail feathers have notched tips. The pink-to-orange lower mandible color is gradually lost with age. The postjuvenal molt is complete and occurs between February and July (Marchant and Higgins 1990).

Adult Plumage, Both Sexes

The period of time to maturity is unknown, and molts of adults are still unstudied. No obvious sex differences in adult plumages exist. Head and neck feathers are barred with blackish-brown and cream, the area above the eyes and crown more uniformly blackish-brown, and the sides of the head more clearly barred with creamy. The neck is similar to sides of head, and the general plumage of upperparts is also similar. The underparts are pale brownish, with creamy barring, becoming increasingly dark brown toward undertail coverts. Wings and tail are blackish-brown, the upperwing coverts barred as in the body feathers, and the underwing coverts mostly whitish. The iris is dark brown; the bill grayish-black, the upper mandible becoming pearl-gray at tip. The gular lobe (rudimentary in females) is dark gray; the legs and feet are grayish-black (Marchant and Higgins 1990).

IDENTIFICATION (FIG. 33C)

In the Hand

The bird is easily distinguished from all other stifftails by its very large size (wing length is at least 165 millimeters, tail length at least 100 millimeters), its uniformly grayish plumage, and the variably large pendant lobe.

In the Field

The musk duck is easily distinguished from all other Australian waterfowl by its very large size, its "battleship-gray" overall plumage pattern, and its very long tail, which is usually kept slightly cocked, especially in males. The birds very rarely are seen on land; because of their large size, they find it harder to walk than do the smaller stifftails. They simply move about on their bellies, pushing themselves forward in a somewhat seallike manner.

ECOLOGY

Habitats and Densities

The favorite breeding habitat for musk ducks consists of deep, permanent wetlands with dense surrounding vegetation and deep central pools of clear water for foraging and display. The ideal breeding situation is found in inland cattail marshes, but marshes with dense lignum (*Muehlenbeckia*) bushes are also used, as are those with various other kinds of surrounding vegetation such as spikerushes (*Eleocharis*) and tea trees (*Melaleuca.*) Some deep swamps that are surrounded with tea trees (*Lepidosperma*) are also used by musk ducks for some of their activities, but breeding has not been reported from these sites. During winter the birds often gather on rather large, deep bodies of water, including mountain and lowland lakes, reservoirs, coastal lagoons, saltwater estuaries, and coastal bays, where they are often found far from shore. Year-round use occurs in wetlands having both dense marginal vegetation and large expanses of open water. Densities on three important nature reserves were from 0.17 to 0.99 per hectare, averaging 0.53 per hectare (Marchant and Higgins 1990).

Foods and Foraging

The studies of Frith, Braithwaite, and McKean (1969) are highly comprehensive. These included analyses of 875 gizzards (but apparently not the esophageal contents) of musk ducks collected throughout the year in an inland cattail-dominated marsh (Barrenbox Swamp) near Griffith in New South Wales. Of these, 399 stomachs were fully analyzed by both volume and occurrence of food items. These samples consisted of 72.6 percent animal materials and 27.4 percent plant materials, with animal foods occurring in 98 percent of the stomachs and plant materials in 80 percent. Most of the animal foods, and 47.5 percent of the total volume, were insects, of which true bugs (Hemiptera) were the most important. Water boatmen (Corixidae) were the second most important insect group, making up 11.1 percent of the total food volume. Few dipteran larvae such as midge larvae were eaten, and microscopic crustaceans were also rare. However, large crayfish (*Cherax*) were an important source of food, as was a freshwater shrimp (*Caridina*), which collectively composed up to 30 percent of the food volume sampled during some months. Other large animals included tadpoles and adult frogs, large freshwater mussels, various fish up to about 10 centimeters long, and even a few downy ducklings of the Australian white-eyed pochard (*Aythya australis*). Most of the plant materials consumed were from deep-water sites. Of these, cattail was the most important (7.1 percent of food, present in 45.2 percent of the samples), including both seeds obtained from the water and leaf bases stripped from growing plants. The seeds and leaves of pondweeds *(Potamogeton)* composed 5.9 percent of the total food volume and occurred in 25 percent of the samples. Several other deep-water plants were present; the remainder of the material consisted mostly of littoral-zone plants and floating plants.

Young ducklings are largely fed directly (bill to bill) by their mothers; the females dive and bring up the food from deep water. Based on an analysis of 62 ducklings, animal materials composed 85.2 of the total sample volume, as compared with 92.7 percent for "immatures" (apparently partially downy juveniles) and 66 to 69 percent for adults. The young birds ate more dragonfly larvae than did adults, and larger amounts of a bug *(Sphaeroderma)* and a water spider *(Arctosa)*, but consumed no mollusks at all, the shells of which may be too hard for their bills to manage (Frith, Braithwaite, and McKean 1969).

Some information on foraging dive durations was provided in Table 6; surprisingly little additional information of this type exists. The birds are reported to forage in water to at least as deep as 6 meters and to remain submerged for up to 60 seconds. The birds normally forage solitarily but have also been observed feeding as a group, diving and swimming in line, or swimming along a broad front (Marchant and Higgins 1990).

Competitors, Predators, and Symbionts

As mentioned in the account of the Australian blue-billed duck, little dietary overlap occurs between these two species, perhaps in large measure because of their overall size differences and their differences in bill structure and foraging behaviors. Few predators are likely to be able catch and kill adult musk ducks; at least adult males are extremely hardy birds, and birds of all ages are highly adept at escape-diving.

ANNUAL CYCLE

Movements and Migrations

Like the blue-billed duck, this species exhibits regular seasonal movements in southeastern Australia, moving into densely vegetated swamps for breeding and into clear lakes and other nonbreeding wetlands for the nonbreeding (winter) season. In inland New South Wales movement out of the breeding swamps begins in March and April and continues until July. In particular, it is probable that juveniles are forced out of the breeding swamps in autumn and winter elsewhere, probably to the south in the Murray River basin and beyond. Those that survive to sexual maturity (presumably in two to three years or more, at least among males) may then move elsewhere to breed (Frith 1967). Besides these regular movements, there are also dispersals associated with periodic changes in surface-water conditions that allow for temporary breeding opportunities in areas such as the Lake Eyre region or elsewhere in the dry interior of Australia. In spite of this species' seemingly limited flying abilities, especially in adult males, such sites are exploited regularly (Marchant and Higgins 1990).

Molts and Plumages

This aspect has not yet been well studied in the wild or in captivity. However, a few points seem clear. As in other stifftails, there is a complete postjuvenal molt, which occurs between about February and July. There is a complete postbreeding molt in adults, which extends in wild birds from April to June, peaking in May. This molt is completed in May by females, and the wing feathers are dropped simultaneously at that time. The prebreeding molt of adults has been described as partial, occurring from early October to December in females and during December among males, with a replacement of the body feathers only (Marchant and Higgins 1990). Contrarily, Frith (1967) reported that two complete molts of wing and tail feathers occur in wild birds, with the postbreeding molt occurring in April to June and the prenuptial molt beginning between October (females) and December (males). The few birds at the Wildfowl Trust likewise had two complete molts per year, involving two flightless periods and two relatively tailless periods. One molting period occurred in May, and the second during August and September. The primaries took two to three weeks to regrow, and the rectrices required about four weeks or more to attain their full lengths (Carbonell 1983).

Breeding Cycle

The best data on breeding cycles comes from the work of Braithwaite and Frith (1969), who collected 281 adult males, 415 adult females, and 286 young birds, including unfledged ducklings, at Barrenbox Swamp, New South Wales. These authors believed that female musk ducks become sexually mature when a year old, whereas males are unlikely to breed until after their first year, perhaps in part because they cannot effectively compete for territories. Data obtained over a five-year period indicated that breeding occurred regularly between June and December, with a peak between August and October. There was little year-to-year variation in breeding durations, suggesting that fixed seasonal timers, such as photoperiod, may set breeding cycles, rather than proximate factors such as rainfall patterns and associated water-level changes. Evidently, the males return to their breeding swamps in a state of relatively low gonadal development and probably only attain their full reproductive potential when and if they manage to secure breeding territories. Unsuccessful males may thus form a reserve population from which territories can be filled whenever they become vacant. Young females are evidently recruited into the adult population and begin to show gonadal development through their first winter, suggesting that most of them breed the following spring. It is possible that females respond to the autumnal photoperiod changes, or a similar external timer, in such a manner that breeding may subsequently become possible whenever suitable environmental conditions develop. As compared with the seemingly long periods that males may remain sexually active, females may exhibit a relatively short breeding period, presumably limiting their opportunities for laying to that time which is most suitable ecologically. In stable environments such as Barrenbox Swamp this most often will occur during spring, but in years of extensive inland flooding it may also occur during the autumn months.

SOCIAL AND SEXUAL BEHAVIOR

Mating System

The mating system of the musk duck is clearly one of polygynous or promiscuous mating. To a degree that is perhaps unrivaled by any other waterfowl species, its mating strategies seem to have been driven by the pressures of classic intrasexual sexual selection, involving relative male dominance. Associated with these male-dominance mating strategies are such behavioral and morphological correlates as strong sexual dimorphism in adult sizes and masses, great sexual differences in signaling devices and social behavior, no involvement of the male in reproduction beyond fertilization, intense competition and fighting among males in holding territories and attracting females, and sometimes even the development of lek-like assemblages consisting of a dominant male and one or more peripheral subordinates (Johnsgard 1967).

Territoriality

Adult males are highly territorial, the territory being used for male advertisement display and foraging. It is not known whether females also normally nest within the male's territory or whether the same territory is defended from year to year. As with the maccoa duck, the musk duck's territory includes a large area of open water and some frontage of emergent vegetation. The male courtship or self-advertisement displays described below appear to be just as strongly directed toward other males, as implicit threat displays, as they are toward females. Curiously, the display activity of adult males often seems to attract immature males, who gather as close to the adult male as possible, as apparent voyeurs. Quite possibly they then attempt to divert any females, but so long as they themselves do not show any display behavior they are seemingly accepted by the resident male. However, other obviously adult males, with fully developed lobes, are not tolerated on a male's territory, and such intrusion often leads to serious, life-threatening fights between them. Johnsgard (1966) observed one such wing-beating and biting fight that lasted several minutes. It terminated when one male finally escaped from the other and retreated as rapidly and as inconspicuously as possible to shore, where it remained stretched out and immobile on the beach for several minutes.

Courtship and Pair-Bonding Behavior

Male displays have been described on many occasions (e.g., Serventy 1946; Stranger 1961; Lowe 1966; Robinson and Robinson 1970), of which the accounts by Johnsgard (1966) and by Fullagar and Carbonell (1986) are most complete. The latter account was largely based on Carbonell's (1983) observations at the Wildfowl Trust and thus overlaps with information presented here. This analysis was based on the actions of a single male from Western Australia, but the data conform fairly closely with the information obtained by Johnsgard (1966) on wild males in southeastern Australia and with Fullagar's observation of several captive males in southeastern Australia. The origins of these birds is of some interest, because of reported possible interpopulation variations in their displays and vocalizations (Robinson and Robinson 1970).

The *paddle-kick* ("position I" of Serventy 1946) display (Fig. 30A, B) represented 44.3 percent of the total display activities tallied and lasted 0.5 to 1.7 seconds (average of 8, 0.57 second). In this movement the male begins with his tail spread and variously raised above the water, his head forward with bill tilted slightly upwards. He then kicks his feet backwards and sideways, producing a noisy splash of water that is primarily directed posteriorly. The cheeks and upper part of the neck are inflated, and the turgid subgular pouch almost touches the water. Before each kick the wings are lifted to the center of the back but are not flapped. In some cases the two feet are kicked asynchronously, but normally they are synchronous. The paddle kick always begins a display sequence, and the male at the Wildfowl Trust performed from 5 to 64 (average of 18, 35) kicks in such sequence, whereas Johnsgard (1966) observed that wild males performed up to 30 (average of 4, 13) paddle kicks before beginning the second phase of display. He also reported an interdisplay interval of 2.4 to 8.3 seconds (average of 19, 4.1 seconds).

The *plonk-kick* ("position II" of Serventy 1946) display (Fig. 30C, D) constituted 10.8 percent of the total displays tallied, and each kick lasted 0.5 to 0.8 second (average of 5, 0.67 second). This display was performed with the tail feathers spread and held vertically at the instant of kicking. The head, neck, and lobe are held as in the paddle kick, and the same wing movement precedes the plonk kick as the paddle kick. Relative to the paddle kick, the water is splashed more vertically, and a loud "plonk" accompanies each splashing movement, presumably as a result of the webbed feet striking the water. Among wild birds, Johnsgard (1966) and Lowe (1966) both reported that this was the most commonly observed form of kick display, but

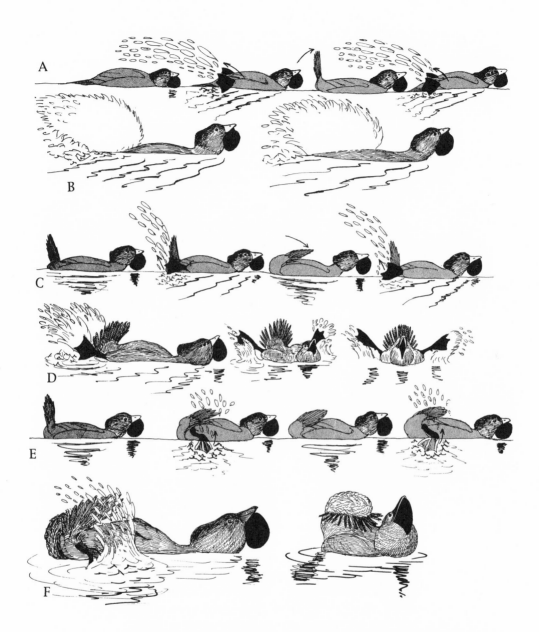

Figure 30. Social-sexual behavior of the male musk duck: (A, B) paddle kick, (C, D) plonk kick, and (E, F) whistle kick. [(A, C, E) After sketches from cine sequences by Carbonell; (B, D, F) based on still photos by Johnsgard.]

at the Wildfowl Trust this was not the case. The maximum number of uninterrupted plonk kicks seen at the Wildfowl Trust was 20, whereas Johnsgard reported up to 51 in wild birds in southeastern Australia. The average interkick interval at the Wildfowl Trust was 6.0 seconds (ranging from 4 to 10), whereas two different estimates of such intervals among wild birds by Johnsgard were both 3.3 seconds (total 134 display intervals). Robinson and Robinson (1970) reported average interkick intervals of 4.72 to 5.64 seconds for birds from Western Australia versus 3.25 seconds for birds from southeastern Australia.

The *whistle-kick* display (Fig. 30E, F) represented 44.9 percent of the total displays tallied and lasted from 0.3 to 0.6 second (average. of 11, 0.57 second). It is associated with the most extreme posture assumed by the male during display, and Johnsgard (1966) found that it also has the greatest time constancy in terms of interdisplay intervals. This averaged 3.7 seconds for 44 intervals timed by him in Australia, with a range of from 2.7 to 4.3 seconds and a standard deviation of 0.28 second, in displays involving several different males. Carbonell reported an interdisplay interval of 3.91 seconds (from 2 to 18) for 204 displays by a single male and a much greater standard deviation of 1.68 seconds. Robinson and Robinson reported average intervals of 4.81 seconds for 37 displays by western birds and 3.37 seconds for 31 displays by eastern birds, with very small standard deviations (0.28 to 0.32 second) in both cases.

During the whistle kick, as well as during intervals between successive kicks, the tail is cocked forward to the point that it presses against the back feathers, fully exposing the undertail coverts. The turgid pouch also touches the water, the neck and head are fully expanded, and the neck is almost entirely submerged. The wings are momentarily lifted just prior to the kick, which tends to send a shower of water laterally on either side of the bird, rather than posteriorly as in the paddle kick and to some degree in the plonk kick. A sharp whistled note is uttered as the feet are kicked.

It is apparent that, rather than being three wholly different displays, these three actions represent an increasing level of display intensity. The only real difference is in the position of the variably cocked tail and the correspondingly different angle of attack of the feet during kicking, the feet being increasingly tilted upward and backward, thus causing the water to spray at varied angles. This display sequence usually progresses from paddle kick to plonk kick (62 percent of 78 observations), and then from plonk kick to whistle kick (37 percent), but once there was a direct transition from paddle kick to whistle kick. Sometimes there are also reversals, as from from plonk kick to paddle kick (65 percent of 52 reversals) and from whistle kick to plonk kick (35 percent).

The vocalizations associated with musk duck displays are a source of great controversy, and apparent differences between vocalizations of eastern and western birds led Robinson and Robinson (1970) to conclude that the two populations have been separated for a long time. Contrary to the views of most earlier authors, Carbonell found that at least in the Wildfowl Trust male from Western Australia that she observed, all three display stages are associated with vocalizations. Thus, all three had a "whit" whistle, which was followed by a "cu" or "whut" note, uttered as a "cu-whit" as the feet were kicked. A "cock" or "grrr" sometimes followed or preceded any of the three kicks, as if the male were trying to clear his throat. However, 28.5 percent of the paddle kicks, 9.0 percent of the plonk kicks, and 2.2 percent of the whistle kicks were performed without the accompanying whistle. This finding is contrary to the statement by Johnsgard (1966) that only the whistle-kick display has an associated and invariable accompanying whistle, a statement based on observations of rather distant wild birds.

Climatic, seasonal, and "emotional" factors have all been considered as possible explanations for the apparent differences in the displays and calls between males from western versus eastern Australia. Yet there is no clear indication that these two populations are fully isolated at present, nor have they necessarily been so isolated in the past. It is possible that age differences among the individual birds studied may in part account for some of the apparent differences that have been observed, in addition to the problems of human interpretation that are inherent in all behavioral descriptions. However, there is good reason to believe that associated male display vocalizations might differ regionally, especially between eastern and western birds (Fullagar and Carbonell 1986). It is clear that knowledge of the musk duck's displays is still incomplete, and a good comparative study of birds from both regions, using the same sound and visual analysis equipment for both populations, would be highly desirable.

Fertilization Behavior

Copulations have been observed and described by only a few observers. Two unpublished accounts summarized by Johnsgard (1965a) suggest that no species precopulatory displays are present; instead the male simply displays with particular intensity when a female approaches closely, seemingly jockeying for position close to her. Finally, the male reportedly throws one wing over the female and pulls her beneath him, so that she is totally submerged during copulation. After both of the observed copulations, the male, with his head submerged, swam rapidly either away from the female or around her. In another account, a female circled a displaying (whistle-kicking) male, while about a half meter away from him, the male rotating to face the circling female. The female then became prone, and the male approached and mounted in the usual manner. In this case the female did not submerge completely, but after treading, the male rushed around the female with his neck outstretched and half submerged, his tail flat on the water. The female sank from view and emerged as the male was completing his circle around her. The two birds then swam off in different directions (Fullagar and Carbonell 1986).

NESTING AND PARENTAL BEHAVIOR

Nests and Nest Building

Musk duck nests are typically well concealed and are usually built by breaking from the center of a cattail clump. Such a clump may be located at the edge of a pool or in an isolated patch, but nests are rarely located far into a cattail thicket. After selecting a site, the bird then weaves the stems into a rough cup that is often so flimsy that the bottoms of the eggs may become wet. Vegetation overhead may be pulled down to provide a partial roof. There may also be a lining of mixed down and leaves, and, in contrast to other stifftails, a substantial amount of down may be added after the clutch is complete and during incubation. Occasionally nests are built on the overhanging branches of tea trees when these touch the water, on the tops of low stumps, or occasionally on the ground, in grass clumps on small islands (Frith 1967). At least in captivity the nest may take about three days to complete, and the same nest may be used for successive clutches. Renesting after clutch failure in the wild is still undocumented, but in captivity as many as four successful clutches have been produced in a single breeding season (Marchant and Higgins 1990).

Egg Laying and Incubation

Egg-laying intervals are still unknown with certainty (between 24 to 98 hours in one case), but clutch-size data are fairly good for wild birds. (See Table 15.) Generally, the clutch is from one to three eggs, but a sufficient number of much larger clutches have been reported to suggest that dump-nesting may be fairly common. Eggs have also been found in the nests of black swans *(Cygnus atratus)* and Pacific black ducks *(Anas superciliosa)*. In captivity, the clutch varies from two to four, with three eggs most commonly reported (Marchant and Higgins 1990).

Even though this is a small clutch size, and the relative clutch mass as compared with that of the adult female is about 21 to 22 percent (Briggs 1988; also Table 16), significant prelaying deposition of body mass occurs. The seasonal increases in mean female weight between the prebreeding and prelaying period was estimated by Briggs (1988) at 24 percent in musk ducks (see weight data above), as compared with an increase of 46 percent in female blue-billed ducks between prebreeding and laying periods. This relatively large, short-term seasonal weight acquisition has been attributed to the short migrations typical of these two species, as compared with those of the more migratory North American ruddy duck, which perhaps cannot "afford" to add so much weight prior to its long spring migration. However, the data of Gray (1980) suggest that very substantial weight gains occur in females of that species (from about 450 to 490 grams to nearly 820 grams, or about a 50 percent increase) after their arrival on the breeding grounds but before laying has begun. At least in captivity, incubation lasts 24 days and probably begins with the laying of the last egg, since the ducklings are known to hatch synchronously.

Hatching and Brood-Related Behavior

In contrast to all other known species of ducks, female musk ducks not only closely attend their young until fledging or even beyond, but even directly feed them bill to bill. Nearly full-grown young may thus be fed; one male weighing 1460 grams, with some down still clinging to his back, was thus seen being fed, as have four-month-old juveniles. Evidently small ducklings are totally dependent on their mothers for food and are unwilling or unable to dive for themselves for some time after hatching. When the female emerges with food for her young, they rush to her with up-tilted bills and loud begging calls, pecking repeatedly at the base of her bill until she feeds them. Interestingly, somewhat similar postures and behavior are used by babies of another curious Australian waterfowl, the magpie goose (*Anseranas semipalmata*) when they are begging for food. The period of time from hatching to fledging is unknown for wild birds but requires three to four months in captivity (Marchant and Higgins 1990). One male raised at the Wildfowl Trust had not completely grown his pouch, nor was he fully displaying, when he was three years old (Carbonell 1983). Other males in captivity have been known to survive for at least as long as 17 years (Fullagar and Carbonell 1986).

REPRODUCTIVE SUCCESS AND STATUS

Relatively little information on reproductive status is available, but the usual brood size is a single duckling. Of 24 class I (all downy) broods, 15 were of single chicks and 9 were of broods with two chicks. Older broods usually consisted of single young, but because of confusion caused by both brood intermingling and unattended young, it was impossible to judge brood size in older ducklings (Braithwaite and Frith 1969). Surveys of ducks on Victoria wetlands made during the summers of 1987 to 1989 indicated that musk ducks then composed less than 1 percent of all ducks counted, with the three-year average being 1480 birds on 477 wetlands. Few birds are shot by hunters, although some are caught in fishing nets. However, prime breeding habitats are increasingly affected by drainage, clearing, grazing, burning, increased salinity, and increased inundation (Marchant and Higgins 1990).

Glossary

Aberrant. Deviating from the normal type.

Active mate choice. Nonrandom mating behavior in which mates are chosen for particular phenotypic traits rather than because of simple proximity or other nonselective ("passive") reasons.

Adaptation. A structural, behavioral, or physiological trait that increases an organism's ability to survive and reproduce.

Advertising behavior. The social behaviors ("displays") of an animal of either sex that serve to identify and broadcast (visually and/or acoustically) its species and sex and may also announce or reveal its relative reproductive capabilities, social status, and general vigor.

Aggression. The component of the hostile interaction spectrum that is associated with threat, fighting, and determining social dominance, the other component being submissive behavior that leads to social subordination.

Agonistic behavior. Hostile interactions between individuals that range from overt attack and fighting to submission and escape—and including both interspecific (often survival-related) and intraspecific (often competition-related) interactions. *See also* Aggression.

Air sac. A structure in the neck region of some grouse, bustards, shorebirds, and ducks that expands that area by air inflation and may also help to modulate or resonate vocal sound production. In some stifftails the neck region is enlarged though esophageal inflation, but such "air sacs" are not directly associated with the avian respiratory system. In others, true tracheal air sacs are present and serve similar visual or sound-related display functions.

Allopatry. Zoogeographic condition wherein populations or species occupy mutually exclusive but sometimes adjacent geographic ranges. Parapatric populations are those in actual extended geographic contact at their boundaries. *See also* Sympatry.

Anatidae. The taxonomic term for the family of ducks, geese, and swans.

Avian. Referring to birds (members of the vertebrate class Aves).

Axillaries. The elongated "armpit" feathers located between the upper flanks and the innermost wing coverts.

Basic plumage. The nonbreeding (winter or, in males, "eclipse") plumage of waterfowl, as opposed to the breeding or "alternate" plumage.

Bimodal. *See* Mode.

Brood. To cover and apply heat to hatched young.

Brood parasitism. Deposition by females of their eggs in the nests of another (sometimes the same) species, leaving them to be incubated and reared by the latter. *See also* Social parasitism.

Bulla. A bony inflation of the syringeal region that may generate, modulate, or resonate vocalizations. *See also* Syrinx.

Call. In ornithology, avian vocalization that is acoustically simple and often innately uttered and perceived, the production of which is usually not limited by sex, age, or season. *See also* Vocalizations.

Character. In biology, a trait or a trait component, such as the use of derived characters in cladistics. *See also* Trait.

Clade. A group of organisms linked by common evolutionary descent; also a branch of a cladogram.

Cladistics. A procedure for arranging taxa by comparing their shared and derived character traits in such a way that the resulting arrangement (or cladogram) reflects their apparent recency of common ancestry.

Clutch. The total number of eggs laid by a single female during an egg-laying cycle and usually (except in social parasites) incubated collectively.

Competition. Supply-and-demand-related interactions between organisms over limited resources, including interactions between individuals or groups (intraspecific competition) as well as between species (interspecific competition).

Congeneric. Belonging to the same genus.

Conspecific. Belonging to the same species.

Convergent evolution. A pattern of evolution during which because of similar ecological adaptations, two or more relatively unrelated lineages progressively approach one another in various phenotypic attributes.

Courtship. A nontechnical term referring collectively to various heterosexual premating behaviors of animals. In monogamous species these usually include diverse "pairing" interactions that are broadly related to forming and maintaining pair bonds, but in polygynous or promiscuous species the behaviors often involve only heterosexual attraction and precopulatory ("mating") signals. *See also* Mating, Pair-forming behavior.

Crepuscular. Pertaining to dawn and twilight periods. *See also* Diurnal, Nocturnal.

Culmen. The upper profile of the bill. Culmen length is the straight-line distance from the tip of the bill to its base.

Dichromatic. Occurring in two discrete color phenotypes, such as in sexual dichromatism, where the adult sexes differ in color (hue or brightness) or pattern.

Dimorphism. Existence of two discrete structural phenotypes in members of a population; used here specifically to refer to grouping based on physically defined distinctions, such as linear mensural differences (size dimorphism) or weight differences (mass dimorphism).

Display. A convenient nontechnical term for a social signal (usually visual or acoustic in birds) that transmits specific information between one individual and another, the latter often but not necessarily of the same species.

Diurnal. Pertaining to the hours of daylight. *See also* Crepuscular, Nocturnal.

Dominance hierarchies. Structured social groupings, either temporary or indefinite, that are organized from the most dominant (alpha) individuals to the least dominant or most subordinate ones.

Dominant. Pertaining to (1) the social superiority of one individual over another (social dominance) or, with reference to a genetic allele, (2) the phenotypic appearance of a heterozygotic individual (genetic dominance). Social dominance may be site-dependent (territorial dominance) or group-related (hierarchical dominance).

Double-brooding. The attempt to raise a second brood after the successful completion of a nesting in the same breeding season.

Dump-nesting. The laying of eggs in a common nest by more than one female.

Eclipse plumage. Dull, often femalelike plumage carried by nonbreeding males of most nontropical duck species; eclipse plumage alternates with the brighter breeding ("nuptial") plumage. In stifftails eclipse plumage may be lacking or not noticeably different from the breeding plumage.

Ecology. The study of the interrelationships between organisms and their biotic and physical environments.

Epigamic. Pertaining to sexual reproduction, especially those structures (secondary sexual characteristics) or behaviors (advertising displays) that serve to facilitate such reproduction.

Ethological characteristics. Behavioral traits, especially those species-specific behaviors that are at least in part genetically controlled, such as those operating in ethological isolation. *See also* Instinct, Isolating mechanisms, Species-specific.

Ethology. The study of behavior from an evolutionary, often naturalistic or comparative, biological viewpoint.

Evolution. Any gradual change; biological evolution involves change as a result of changing gene frequencies between generations.

Extant. Nonextinct.

Extinct. No longer surviving anywhere.

Extirpated. No longer surviving in a particular location.

Extra-pair copulations. Additional copulations obtained by an already paired individual with a conspecific other than its own mate. Sometimes termed "sneaky" copulations. *See also* Forced copulations.

Facultative polygyny/promiscuity. A mating system whereby, under certain conditions, males of a species that is normally monogamous fertilize additional females in a single breeding season. *See also* Extra-pair copulations.

Fecundity. A measure of reproductive potential, especially with regard to females.

Female-choice behavior. The variable attraction of females to individual or grouped males for mating, thus providing a basis for intersexual selection.

Female-defense polygyny. A mating system in which some males are able to gain and control sexual access to two or more breeding females simultaneously; may be "economically" or "noneconomically" based—the former if sexually successful males also defend resource-rich territories. Sometimes called harem polygyny. *See also* Male-dominance polygyny, Resource-defense polygyny.

Feral. Living in a secondarily wild state following a period of captivity or domestication.

Fertility. A measure of reproductive potential, especially with regard to males. *See also* Fecundity.

Fertilization. The process by which two gametes unite to form a zygote; in birds internal fertilization is achieved by copulation.

Fledge. To take flight for the first time; the fledging period is the interval between hatching and fledging.

Flightless period. The period during molting when the flight feathers are being replaced and thus flight is impossible.

Foraging niche. Food-related adaptations of a species, including not only a species' foods but also its behavioral characteristics affecting foraging efficiency.

Forced copulations. Rapelike behavior in birds and other nonhumans. *See also* Extra-pair copulations.

Fundamental frequency. The lowest sound frequency in a group of harmonically related sounds. *See also* Harmonic.

Gamete. A haploid germ cell (egg or sperm).

Generic. Pertaining to a genus.

Genotype. The genetic identity of a gene locus of an individual; more broadly speaking, the gene constitution of an individual.

Genus (pl. genera). The taxomic category immediately above the species level; also the first component of an organism's binomial name.

Habitat. The specific environment within which a particular organism lives, including both its physical and biotic components. *See also* Niche.

Harem. A group of females simultaneously under the sexual control of a single male, as in harem polygyny. *See also* Female-defense polygyny.

Harmonic. One in a series of consecutive multiple frequencies, or overtones, variably resonated above a fundamental sound frequency. *See also* Fundamental frequency.

Hectare. An metrically defined area (10,000 square meters) equal to 2.47 acres.

Heterosexual. Pertaining to between-sex (intersexual) interactions or differences.

Hierarchy. A classification of individuals according to relative dominance.

Homologous. Pertaining to structural, physiological, or behavioral traits that are shared by two or more groups and reflect their common evolutionary ancestry.

Hybridization. The interbreeding of individuals from two genetically unlike natural populations, such as different races (intraspecific hybridization), different species (interspecific hybridization), or different genera.

Inciting. A mate-choice and pair-bonding behavior typical of many duck species, during which females seemingly encourage their mates or prospective mates to threaten or attack other individuals, often other conspecific males.

Incubation period. The interval between the initial application of heat to an egg and its hatching.

Individual distance. The distance within which one individual will not tolerate the presence of another; such spacing behavior is especially evident in flocking or colonial species that do not use territoriality for spacing out.

Innate. Pertaining to genetically transmitted traits.

Instinct. A genetically inherent (innate) behavioral trait that appears to be more complex than various simpler innate responses (such as reflexes or taxes) and whose performance seemingly depends on specific and transitory internal states ("tendencies" or "drives").

Interacetabular width. The distance between the dorsal edges of the paired femur sockets of the pelvic girdle.

Intergeneric. Pertaining to interactions between genera.

Interscapulars. Back feathers lying between the scapulars.

Intersexual. Pertaining to heterosexual interactions.

Interspecific. Pertaining to interactions between species.

Intrageneric. Pertaining to interactions between species of a single genus.

Intraspecific. Pertaining to within-species interactions.

Invertebrate. Any nonvertebrate animal; an animal lacking a backbone.

Iridescent. Producing rainbowlike hues through light refraction and reflection, as typical of many duck speculum areas.

Isolating mechanisms. Inherent ("intrinsic") characteristics of individuals of a species that tend to prevent successful interbreeding with other sympatric species. *See also* Reproductive isolation.

Juvenile. A young but fledged bird that still retains part or all of its juvenal plumage.

Lekking. A category of epigamic behavior in which males gather in variably clustered arenas (leks) for reproductive purposes.

Male-dominance polygyny. A "noneconomic" polygynous mating system in which male reproductive success is related to individual social dominance and associated sexual access to females, rather than to defense of a resource-rich territory.

Mass ratio. Used here to indicate the average adult male-to-female weight ratio, expressed relative to unity (e.g., 1.5:1).

Mate-choice behavior. The intersexual aspect of sexual selection; namely, making mating-partner choices based on individual heterosexual attraction attributes.

Mate competition. The direct and indirect competition among members of one sex for sexual access to the other sex.

Mating. An term variously used with reference either (1) to initial pair bonding (pairing behavior) or more specifically (2) to copulation.

Mating strategy. A species' evolved behavioral basis for engaging in a particular mating system. *See also* Mating system, Strategy.

Mating success. A measure of an individual's capacity for transmitting its genes to the next generation.

Mating system. A general descriptive term for various kinds of mating strategies, such as monogamy, polygyny, polyandry, and promiscuity. Definitions sometimes also include additional mating attributes, such as the duration of pair bonding (if any), the degree of random or selective mating, and possible differential parental investments. *See also* Mating.

Mean. A numerical average.

Mechanical sounds. Nonvocal (nonsyringeal) noises produced by percussion, stridulation, feather vibration, or other similar means.

Mobile arena (or lek). An assemblage of males who are competing sexually for females and whose interactions are not dependent upon specific territories nor restricted to fixed display sites.

Mode. The most frequent item or value in a series of observations.

Modulation. Controlled temporal variations in sound frequencies or amplitudes, especially vocalizations. *See also* Resonance.

Molting. In birds, the sequential loss and replacement of individual feathers or entire plumages.

Monogamy. A mating system in which adults of both sexes establish pair bonds that persist through at least one breeding cycle (single-brood monogamy) or for a longer, sometimes lifelong, period (permanent or indefinite monogamy). Single-brood monogamous species sometimes become facultatively polygamous if the breeding season is long enough.

Monomorphism. Existence of a single morphological phenotype ("morph") in members of a population; in a monomorphic species the sexes appear phenotypically identical as adults. *See also* Morphism.

Monotypic. Descriptive of a taxonomic group having a single representative, such as a monotypic genus with a single included species.

Morph. An example of any of the phenotypic variants that occur in cases of sexual or nonsexual morphism. Nonsexual morphs are sometimes called phases.

Morphism. Here used to refer to structural or behavioral phenotypic variations in a population's traits that result in dimorphism or polymorphism. *See also* Dimorphism, Monomorphism, Polymorphism, Sexual dimorphism.

Morphological. Pertaining to structures, especially external or generally apparent structures as opposed to internal (anatomical) structures, of an organism.

Mortality rate. The rate at which individuals are removed from a population by death; often calculated on an annual basis and usually expressed as a percentage (e.g., 15 percent); reciprocally related to the population's survival rate, the sum of the two being 100 percent (e.g., $M = 15\%$, $S = 85\%$). *See also* Turnover rate.

Natural selection. In a contemporary and broad sense, the changes in gene frequencies and associated traits resulting from differential survival and reproduction of individuals within an interbreeding population.

Nearctic region. A zoogeographic region encompassing nontropical portions of North America.

Niche. The behavioral, morphological, and physiological adaptations of a species to its habitat; also sometimes defined from an environmental standpoint—namely, the range of ecological conditions under which a species potentially exists (fundamental niche), in which it survives best (preferred niche), or in which it actually survives (realized niche).

Niche segregation. The ecological separation of two populations (or subpopulations) so as to reduce ecological competition between them.

Nocturnal. Pertaining to the nighttime hours. *See also* Diurnal.

Nuptial. Pertaining to breeding, as in nuptial plumages or nuptial displays.

Nuptial plumage. The breeding plumage of a bird; sometimes also called the "alternate" plumage.

Ontogeny. The temporal development of an individual, from zygote to adult. *See also* Phylogeny.

Operational sex ratio. The ratio of fertilizable females available to sexually active males at any given time.

Pair bond. A variably prolonged and individualized association between an adult male and an adult female that is established sometime before breeding.

Pair-forming behavior. The means by which pair bonds are formed; also known as "courtship" behavior.

Palearctic region. A zoogeographic region encompassing all of Europe, Asia south to the Himalayas, and Africa north of the Sahara.

Parapatry. *See* Allopatry.

Parasitic nesting. The laying of one or more eggs in the nest of another bird of the same or different species, with no further attempt at their incubation; not always easily distinguished from dump-nesting.

Parental investment. Any investment (food, energy, or time) made by the parent in an individual offspring that increases the offspring's chance of surviving at the cost of the parent's ability to invest in other activities.

Phenotype. The collective characteristics composing the attributes of a individual organism and resulting from interactions between the organism's genotype and the environment.

Phyletic dendrogram. A treelike representation of a group's evolutionary history.

Phylogenetic order. A linear sequence of taxonomic groups that best reflects their known phyletic affinities.

Phylogeny. The evolutionary history of an organism or related group of organisms. *See also* Ontogeny.

Plumage. A group of feathers associated with a single molting period and thus representing a single feather generation.

Pochards. A collective vernacular name for species of the diving-duck genus *Aythya*.

Polygamy. A mating system in which either individual males gain sexual access to more than one female (polygyny) or individual females gain sexual access to more than one male (polyandry). *See also* Polygyny, Promiscuity.

Polygyny. A mating system in which males mate with two or more females during a single reproductive cycle, either establishing multiple pair bonds simultaneously or pairing with several different females in rapid sequential, sometimes overlapping, succession.

Polymorphism. The presence of two or more genetic types ("morphs") in a population.

Postacetabular. Located behind the acetabulum (femur socket).

Precocial. Descriptive of those animals (here specifically birds) that at or shortly after hatching are feathered sufficiently (ptilopaedic) to retain body heat, that often leave the nest site well before fledging (nidifugous), and that are often sufficiently developed to soon begin foraging for themselves.

Primaries. Major flight feathers of birds that are supported by the hand and finger bones. *See also* Secondaries.

Primitive. Pertaining to generalized traits that appeared early in a group's phylogeny but later gave rise to derived (but not necessarily more complex) traits.

Promiscuity. A polygamous mating system in which breeding adults form no heterosexual pair bonds but instead associate with the opposite sex only for reproductive purposes; not always easily distinguished from polyandry and polygyny, which (at least in birds) imply the existence of temporary pair bonding.

Race. *See* Subspecies.

Rape. *See* Forced copulations.

Rectrices. The tail feathers of birds. *See also* Tail coverts.

Releaser. *See* Sign-stimulus.

Reproductive isolation. The genetically based isolation of one species from another by various devices or "isolating mechanisms" such as behavioral (ethological mechanisms), time-dependent (temporal mechanisms), structural (mechanical mechanisms), and habitat-related (ecological mechanisms) differences.

Resonation. The transmission and amplification of vocalizations by nonsyringeal structures, especially those having particular frequency characteristics. *See also* Air sac, Tracheal air sac.

Resource. An environmental attribute needed for an organism's survival and reproduction, including but not necessarily restricted to those that are in limited supply and over which competition might occur.

Resource-defense polygyny. A polygynous mating system in which the male's territory includes environmental resources of potential use to the females with whom he breeds, thus composing an "economic" mating system.

Ritualized. In ethology, descriptive of behaviors that have gradually acquired signal functions (become displays) through evolution, the process being called ritualization. *See also* Display, Signaling behavior.

Salt glands. Excretory glands located above the eyes that remove excess salts (through the nostrils) among birds that regularly drink salty water.

Scapulars. The "shoulder" feathers, located between the median upperwing coverts and the back.

Scientific name. The Latin (or latinized) name of an organism, typically consisting of two elements (binomial), an initial genus or generic name (which is always capitalized) followed by a species or specific name (never capitalized). Three-part names are used whenever two or more geographic races (subspecies) of the species are individually recognized.

Scutellated tarsus. The somewhat rectangular tarsal plates aligned in a vertical series down the front of the lower leg, as opposed to being more rounded or hexagonal and arranged in a weblike (reticulated) pattern.

Secondaries. Major flight feathers that are associated with the forearm (radius and ulna) of birds. *See also* Primaries.

Secondary sexual characteristic. Any sex-related trait (usually limited to or best developed in adults of only one sex) associated with obtaining a mate or facilitating reproduction but usually not directly required for such reproduction or parental care—the latter structures being primary sexual characteristics.

Sex ratio. The ratio of males to females in a population, ranging from the ratio existing at fertilization (primary sex ratio) to that of mature adults (tertiary sex ratio), and shown either as a direct male-to-female ratio (e.g., 1.5:1) or as a proportional percentage (e.g., 60:40).

Sexual dimorphism. Used broadly, the condition by which the adults of a bird species variously exhibit measurably different secondary sex characteristics, including general external morphology or linear dimensions (size dimorphism); overall adult body weights (mass dimorphism); and differences in hue, brightness, or patterns of feather and/or softpart phenotypes (sexual dichromatism).

Sexual selection. A type of natural selection in which the evolution and maintenance of traits of one sex result from the social interactions that produce differential individual reproductive success.

Shelducks. A collective vernacular name for ducklike species (*Tadorna* spp.) of the waterfowl tribe Tadornini. Shelducks and the more gooselike sheldgeese represent a structural and ethological transition between true geese (Anserinae) and more typical ducks (Anatinae).

Signaling behavior. Communication behavior during which information is transmitted from one individual to another through the production of social signals ("displays") such as posturing, vocalizations, or other channels of transmission. *See also* Display, Signaling device.

Signaling device. A structure (such as feathers or softparts) that facilitates the production and broadcasting of social signals ("displays" or sign-stimuli) by an individual. *See also* Display, Sign-stimulus.

Sign-stimulus. In traditional ethological terminology, a signal (visual or acoustic) produced by one individual that "releases" a instinctive response ("fixed action pattern") in another individual who is already in an appropriate internal state conducive to performing such a response (i.e., having an adequate "specific action potential"). The "releaser" is that part of the overall sign-stimulus that is most important in transmitting the signal. *See also* Display.

Single-brood monogamy. Monogamous mating systems in which a pair bond persists through the completion of a single nesting cycle and in which both sexes typically participate to some degree in parental responsibilities. *See also* Monogamy.

Social dominance. *See* Dominance hierarchies, Hierarchy.

Social parasitism. Broadly speaking, exploitive intraspecific or interspecific behavior such as brood parasitism and various kinds of piracy, including food stealing (kleptoparasitism), copulation stealing (cuckoldry), and stealing of nest sites. *See also* Brood parasitism, Extra-pair copulations.

Speciation. The development of reproductive (intrinsic) isolation by a population or group of populations over time; also called species proliferation or species formation.

Species. Biologically, a population or populations of actually or potentially interbreeding individuals that are reproductively isolated from all other populations. The term *species* (unchanged in the plural) also refers to the taxonomic category between the genus and the subspecies; it is written as the second and subsidiary (uncapitalized) component of a two-part (binomial) name, with the generic component taking precedence and being capitalized.

Species-recognition behavior. The exchange of signals between individuals that serves to transmit information concerning species identification; such signals are especially important during intraspecific courtship (epigamic species recognition), but they sometimes also function interspecifically, between competing species (agonistic species recognition).

Species-specific. Pertaining to those traits typical of and also unique to a species, especially traits concerned with reproductive isolation.

Species-typical. Pertaining to traits that are typical of but not necessarily unique to a species. Other taxon-typical traits may be characteristic of a genus, family, etc.

Speculum. A brightly patterned, often iridescent area of feathers on the wings of some birds, especially ducks.

Stereotyped. Descriptive of behavior patterns having marked rigidity in form, duration, and/or amplitudinal aspects, presumably reflecting their innate bases.

Stifftails. A collective vernacular name for various long-tailed diving ducks of the tribe Oxyurini; "typical stifftails" comprise the genera *Oxyura* and *Biziura*.

Strategy. A set of behaviors a species or individual may follow in order to maximize its fitness.

Subfamily. A major taxonomic subdivision of a family, such as the duck subfamily Anatinae.

Subordinate. A socially inferior or submissive individual in a dominance hierarchy.

Subspecies. One or more populations of a species that can be geographically and morphologically distinguished from other such populations and that have gene and genotype frequencies differing from other such subspecies; subspecies are also often called races.

Superspecies. A group of closely related, entirely or essentially allopatric populations that are too distinct taxonomically to be considered as a single species.

Sympatry. The simultaneous occurrence of two or more populations in the same area, especially during the breeding season; often used as a criterion for biologically determining reproductive isolation and defining completed speciation. *See also* Allopatry.

Syrinx (pl. syringes). The vocal organ of birds, functionally comparable to the larynx of mammals. The syrinx is responsible for all avian vocalizations, which may be modulated or resonated, however, by other structures such as the trachea and anterior esophagus. *See also* Air sac, Bulla, Vocalizations.

Tail coverts. Shorter feathers covering the dorsal (uppertail coverts) and ventral (undertail coverts) bases of the tail feathers (rectrices).

Taxon (pl. taxa). A group of organisms representing a particular category of biological classification, such as the genus *Anas*.

Taxonomy. A system of naming and describing groups of organisms (taxa) and of organizing such groups in ways that reflect established phyletic relationships.

Temporal. Pertaining to time.

Territory. An area defended by an animal against others of its species, and also occasionally against other species, and within which it tends to be socially dominant. Territories may be defended only during the breeding season (breeding territories), continuously (permanent territories), or for other seasonal periods (such as winter territories) and may include part or all of the individual's home range. Moving or "revolving" territories (those centering around mates or families) also exist.

Tertiary sex ratio. The ratio of adult males to females in a population. *See also* Operational sex ratio.

Tracheal air sac. Inflatable structure that is directly attached to the tracheal tube (windpipe) rather than being an outgrowth of the lungs. *See also* Air sac.

Trait. A measurable phenotypic attribute (behavioral, structural, or physiological), especially one that is at least in part genetically controlled. Traits may have several character components. *See also* Character.

Tribe. A taxonomic category positioned between the genus and subfamily, such as the stifftail subfamily Oxyurini.

Turnover rate. The period of time needed to replace an entire age class within a population.

Vernacular name. The "common" English name of an organism (usually a species), as opposed to its scientific or Latin name.

Vocalizations. Sounds, including both calls and songs, generated (at least among birds) by the syrinx and sometimes modulated or resonated by nonsyringeal structures. *See also* Fundamental frequency, Harmonic, Modulation, Resonation, Syrinx.

Waterfowl. A collective vernacular name for ducks, geese, and swans of the family Anatidae. Called wildfowl in Britain.

Wing coverts. Feathers of a bird's wings other than its large flight feathers. *See also* Primaries, Secondaries.

Wing loading. A measure of a bird's overall mass relative to its wing surface area; higher wing loadings require greater energy expenditure for flight. A similar "buoyancy index" provides a better estimate of energy requirements when comparing birds of markedly differing masses; higher buoyancy indices reflect lower wing loadings.

A P P E N D I X

Key to In-Hand Species Identification of Stiff-Tailed Ducks and the Freckled Duck

To use this "dichotomous key," choose first between the paired initial descriptors A and AA the one that fits the unknown species best. Then choose between B and BB, C and CC, etc., until a species name is reached.

    A. Tail barely differentiated from or longer than tail coverts, the rectrices neither distinctly narrowed nor stiffened.

      B. Culmen 46 to 56 millimeter long; bill much flattened toward tip: *Stictonetta,* freckled duck, only species (Fig. 31A).

    BB. Culmen maximum 46 millimeter; bill tip not flattened: *Heteronetta,* black-headed duck, only species (Fig. 31B).

   AA. Tail much longer than tail coverts, the rectrices very narrow, with stiffened shafts and sharply pointed tips.

      B. Up to 18 long (50 to 100 millimeter) rectrices present; bill no more than 25 millimeter wide at base: *Oxyura* (six extant species)

        C. Culmen 30 to 35 millimeter; wings with a large white speculum (wing patch): subgenus *Nomonyx,* masked duck, only species (Fig. 31C)

      CC. Culmen 36 to 48 millimeter; wing entirely dark brown above: subgenus *Oxyura*

        D. Bill distinctly swollen basally, culmen 42 to 48 millimeter.

        E. Bill moderately swollen; tail maximum 75 millimeter; head entirely blackish (males) or flanks coarsely patterned with dark brown (females and immatures): maccoa duck (Fig. 32B).

EE. Bill highly swollen; tail over 75 millimeter; head mostly white (adult males) or flanks finely patterned with dark brown (females and immatures): white-headed duck (Fig. 32C).

DD. Bill only slightly swollen basally, culmen 36 to 42 millimeter.

  E. Maximum width of bill usually over 22 millimeter; males often with white on cheeks: ruddy duck (North American race, Fig. 32A).

EE. Maximum width of bill usually under 22 millimeter; males never with white on cheeks.

   F. Culmen usually under 40 millimeter; females and immatures with obscure facial pattern: Argentine blue-billed duck (Fig. 33A).

  FF. Culmen usually over 40 millimeter; females and immatures with strongly contrasting facial pattern: Australian blue-billed duck (Fig. 33B).

BB. With 20 to 24 very long (100 to 120 millimeter) rectrices present; bill more than 30 millimeter wide at base: *Biziura,* musk duck, only extant species (Fig. 33C),

Figure 32. Adult male, female, and downy young of the (A) North American ruddy duck, (B) maccoa duck, and (C) white-headed duck.

Figure 33. Adult male, female, and downy young of the (A) Argentine blue-billed duck, (B) Australian blue-billed duck, and (C) musk duck.

References

Because of the relatively scanty world literature available on the stiff-tailed ducks, this list of references includes a number of sources that are not specifically cited in the text. By way of explanation, the journal *Oxyura* is irregularly published in Spain (Apdo. de Correos 3.059, Cordoba) by the Amigos de la Malvasia Association; the journal is largely devoted to studies on the biology of the white-headed duck and its associated wetland habitats in Spain.

Adams, J., and E. R. Slavid. 1984. Cheek plumage in the Colombian ruddy duck *Oxyura jamaicensis*. *Ibis* 126:405–7.

Ali, S., and S. D. Ripley. 1968. *Handbook of the birds of India and Pakistan, together with those of Nepal, Sikkim, Bhutan and Ceylon*. Vol. 1. Oxford Univ. Press, Bombay.

Alisauskas, R. T., and C. D. Ankney. 1994. Costs and rates of egg formation in ruddy ducks. *Condor* 96:11–18.

Amat, J. A. 1982. The nesting biology of ducks in the marismas of the Guadalquivir, South-West Spain. *Wildfowl* 33:94–104.

———, 1984. Interacciones entre los patos buceadores en una laguna meridional española. Donaña, *Acta. Vert.* 11:105–23.

———, and A. Sanchez. 1982. Biologia y ecologia de la malvasia *(Oxyura leucocephala)* en Andalucia. Donaña, *Acta. Vert.* 9:231–320.

Anon. 1993. Summary & recommendations, International *Oxyura jamaicensis* workshop, 1–2 March 1993, Arundel, U.K.

Anstey, S. 1989a. Progress in the white-headed duck *Oxyura leucocephala* action plan project. *Wildfowl* 40:141.

———. 1989b. The status and conservation of the white-headed duck *Oxyura leucocephala*. IWRB Spec. Pub. 10. IWRB, Slimbridge, U.K.

Arenas, R., and J. A. Torres. 1992. Biología y situacíon de la malvas a en España. *Quercus* 73:14–21.

Attiwill, A. R., J. M. Bourne, and S. A. Parker. 1981. Possible nest parasitism in the Australian stiff-tailed ducks (Anatidae, Oxyurini). *Emu* 81:41–42.

Barattini, L. P., and R. Escalante. 1971. *Catalogo de las aves Uruguaya.* Pt. 2, *Anseriformes.* Intendencial Municipal de Montevedeo, Uruguay.

Barcelona, F. R. 1976. Ruddy duck food habits in northwestern California. M.S. thesis, Humboldt State Univ., Arcata, Calif.

Beard, E. B. 1964. Duck brood behavior at the Seney National Wildlife Refuge. *J. Wildlife Mgmt.* 28:492–521.

Beddard, F. E. 1898. *The structure and classification of birds.* Longmans, Green & Co., London.

Bellrose, F. C. 1980. *Ducks, geese and swans of North America.* 3d ed. Wildlife Mgmt. Inst., Washington, D.C.

———, T. G. Scott, A. S. Hawkins, and J. B. Low. 1961. Sex ratios and age ratios in North American ducks. *Ill. Natl. Hist. Surv. Bull.* 27:391–674.

Belton, W. 1984. Birds of Rio Grande do Sul, Brazil. Pt. 1. *Bull. Am. Mus. Nat. Hist.* 178:1–631.

Bennett, L. J. 1938. *The blue-winged teal, its ecology and management.* Collegiate Press, Ames, Iowa.

Bergan, J. F. 1986. Aggression and habitat segregation among diving ducks wintering in South Carolina. M.S. thesis, Texas Tech Univ., Lubbock.

———, and L. M. Smith. 1989. Differential habitat use by diving ducks wintering in South Carolina. *J. Wildlife Mgmt.* 53:1117–26.

———, ———, and J. J. Mayer. 1989. Time-activity budgets of diving ducks wintering in North Carolina. *J. Wildlife Mgmt.* 43:769–76.

Blackburn, T. M. 1991. The interspecific relationship between egg size and clutch size in waterfowl. *Auk* 108:209–11.

Bond, J. 1958. *Third supplement to the checklist of the birds of the West Indies.* Acad. Nat. Sci., Philadelphia.

———. 1961. *Sixth supplement to the checklist of the birds of the West Indies.* Acad. Natl. Sci., Philadelphia.

Bottjer, P. D. 1983. Systematic relationships among the Anatidae: An immunological study with a history of anatid classification, and a system of classification. Ph.D. diss., Yale Univ., New Haven, Conn.

Bourne, J. M., and S. A. Parker. 1981. Possible nest-parasitism in the Australian stiff-tailed ducks (Anatidae, Oxyurini). *Emu* 81:41–42.

Braithwaite, L. W. 1976. Notes on the breeding of the freckled duck in the Lachland River valley. *Emu* 76:127–32.

———, and H. J. Frith. 1969. Waterfowl in an inland swamp in New South Wales. I. Breeding. *CSIRO Wildlife Res.* 14:65–109.

Briggs, S. V. 1988. Weight changes and reproduction in female blue-billed and musk ducks, compared with North American ruddy ducks. *Wildfowl* 39:98–101.

Brown, A. G. 1949. Display of blue-billed ducks. *Emu* 48:315.

Brown, L. H., E. K. Urban, and K. Newman (eds.). 1982. *The birds of Africa.* Vol. 1 (Ostrich to Hawks). Academic Press, N.Y.

Brush, A. H. 1976. Waterfowl feather proteins: Analysis of use in taxonomic studies. *J. Zoology* (London) 179:467–98.

Callaghan, D. A., and A. J. Green. 1993. Wildfowl at risk. *Wildfowl* 44:149–69.

Campbell, A. J. 1899. On the trachea of the freckled duck of Australia *(Stictonetta naevosa). Ibis,* ser. 7 5:362–64.

Carbonell, M. 1983. Comparative studies of stiff-tailed ducks (Tribe Oxyurini, Anatidae). Ph.D. diss., Univ. College, Cardiff, Wales.

Casos, A. E. 1992. Nidificaíon simpatríca de los patos zambuillidores *Oxyura ferruginea* y *O. vittata* en la Argentina. *Nuestras Aves* 27:33.

Clancey, P. A. 1967. *Gamebirds of southern Africa.* Purnell & Sons, Cape Town.

Clapp, R. B., D. Morgan-Jacob, and R. C. Banks. 1982. *Marine birds of the southeastern United States and Gulf of Mexico.* Part II, *Anseriformes.* U.S. Dept. Interior, Biological Services Prog., FSW/OBS–82/20.

Clark, A. 1964. The maccoa duck (*Oxyura maccoa* [Eyton]). *Ostrich* 35:264–76.

———. 1969. The behavior of the white-backed duck. *Wildfowl* 20:71–74.

———. 1978. Maccoa duck displays. *Ostrich* 49:86.

Collar, N. J., and P. Andrew. 1988. *Birds to watch. The ICBP world checklist of threatened birds.* Smithsonian Inst. Press, Washington, D.C.

Cook, P. A., W. R. Siegfried, and P. G. H. Frost. 1977. Some physiological and biochemical adaptations to diving in three species of ducks. *Comp. Biochem. Physiol.* (A) 57:227–28.

Cottam, C. 1939. Food habits of North American diving ducks. *U.S. Dept. Agric. Tech. Bull.* 643:1–140.

Cotter, W. B., Jr. 1957. A serological analysis of some anatid classifications. *Wilson Bull.* 69:291–300.

Cramp, S., and K. E. L. Simmons (eds.). 1977. *Handbook of the birds of Europe, the Middle East and North Africa.* Vol. 1 (Ostrich to Ducks). Oxford Univ. Press, Oxford.

Daguerre, J. B. 1923. Parasitismo del pato *Heteronetta atricapilla. Hornero* 3:194–95.

Dean, R. J. 1970. *Anas hottentota* and *Oxyura maccoa* eggs in one nest. *Ostrich* 41:216.

del Hoyo, J., A. Elliott, and J. Sargatal (eds.). 1992. *Handbook of the birds of the world.* Vol. 1. Lynx Edicions, Barcelona.

Delacour, J. 1954–64. *The waterfowl of the world.* 4 vols. Country Life, London. (Reprinted in 1973 by Arco Pub. Co., N.Y.)

———, and E. Mayr. 1945. The family Anatidae. *Wilson Bull.* 56:3–55.

Delnicki, D. 1975. The masked duck. *Ducks Unlimited* 41:46–60.

Dementiev, G. P., and N. A. Gladkov (eds.). 1952. *Birds of the Soviet Union.* Vol. 4. Moscow. (English translation, Jerusalem, 1967)

Downer, A., and R. Sutton. 1990. *Birds of Jamaica: A photographic field guide.* Cambridge Univ. Press, Cambridge.

Duran, A. 1961. Notes on keeping the white-headed duck, *Oxyura leucocephala,* in captivity. *Avicult. Mag.* 67:160–61.

Euliss, N. H. 1989. Assessment of drainwater evaporation ponds as waterfowl habitat in the southern San Joaquin Valley, California. Ph.D. diss., Oregon State Univ., Corvallis.

Evans, C., A. Hawkins, and W. Marshall. 1952. Movements of waterfowl broods in Manitoba. U.S. Fish & Wildlife Serv. *Spec. Sci. Rept.: Wildlife* 16:1–47.

Eyton, T. C. 1838. *A monograph of the Anatidae, or duck tribe.* London.

Faith, D. W. 1989. Homoplasty as pattern: Multivariate analysis of morphological convergence in anseriforms. *Cladistics* 5:235–58.

Farner, D. S. 1955. Birdbanding in the study of population dynamics. In *Recent studies in avian biology,* ed. A. Wolfson, 392–449. Univ. Ill. Press, Urbana.

Featherstone, J. D. 1975. Aspects of nest site selection in three species of ducks. Ph.D. diss., Univ. Toronto, Ontario.

ffrench, R. 1991. *Field guide to the birds of Trinidad and Tobago.* 2d ed. Cornell Univ. Press, Ithaca.

Fjeldså, J. 1986. Color variation in the ruddy duck (*Oxyura jamaicensis andina*). *Wilson Bull.* 98:592–94.

———, and N. Krabbe. 1990. *Birds of the high Andes.* Svendbord, Denmark.

Forbes, W. A. 1882. Notes on some points in the anatomy of an Australian duck (*Biziura lobata*). *Proc. Zool. Soc. London,* 455–58.

Frith, H. J. 1964. Taxonomic relationships of *Stictonetta naevosa*. *Nature* 202: 1352.

———. 1965. Ecology of the freckled duck, *Stictonetta naevosa*. *Emu* 64:42–47.

———. 1967. *Waterfowl in Australia*. East-West Center Press, Honolulu.

———, L. W. Braithwaite, and J. L. McKean. 1969. Waterfowl in an inland swamp in New South Wales. II. Food. *CSIRO Wildlife Res.* 14:17–64.

Fullagar, P. J., and M. Carbonell. 1986. The display postures of the musk duck. *Wildfowl* 37:142–50.

———, C. C. Davey, and D. K. Rushton. 1990. Social behaviour of the freckled duck *Stictonetta naevosa* with particular reference to the axel-grind. *Wildfowl* 41:52–61.

Gomez-Dallmeier, F., and A. Cringan. 1990. *Biology, conservation and management of waterfowl in Venezuela*. Editorial ex Libris, Caracas.

Gooders, J., and T. Boyer. 1986. *Ducks in Britain and the Northern Hemisphere*. Facts on File, N.Y.

Gray, B. J. 1980. Reproduction, energetics, and social structure of the ruddy duck. Ph.D. diss., Univ. Calif., Davis.

Green, A. J., and S. Anstey. 1992. The status of the white-headed duck *Oxyura leucocephala*. *Bird Cons. Internat.*, 2:185–200.

———, and B. Hughes. *Oxyura* crisis. *Internat. Waterfowl & Wetlands Res. Bur. Newsletter* 8:18.

Hartman, F. A. 1955. Heart weights in birds. *Condor* 57:221–38.

———. 1961. Locomotory mechanisms in birds. *Smithsonian Misc. Coll.* 143: 1–91.

Haverschmidt, F. 1958. *Birds of Surinam*. Oliver & Boyd, London.

———. 1972. Bird records from Surinam. *Bull. Brit. Orn. Club* 92:49–53.

Hays, H., and Habermann, H. M. 1969. Note on the bill color of the ruddy duck, *Oxyura jamaicensis rubida*. *Auk* 86:765–66.

Heintzelman, D., and C. J. Newberry. 1964. Some waterfowl diving times. *Wilson Bull.* 76:291.

Hellmayr, C. E., and B. Conover. 1948. Catalogue of birds of the Americas. Field Mus. Nat. Hist., *Zool. Ser.* 13 (Pt. 1, No. 2):1–434.

Hewston, N. 1992. The North American ruddy duck in Europe—A new threat to the white-headed duck. *Intern. Wild Waterfowl Assn. Newsletter*, Fall 1992, 8–10.

Hillgarth, N., and J. Kear. 1982. Diseases of stiff-tailed ducks in captivity. *Wildfowl* 33:140–44.

Hinde, R. A. 1955–56. A comparative study of the courtship of certain finches (Fringillidae). *Ibis* 97:706–45; 98:1–23.

Hochbaum, H. A. 1939. Waterfowl studies at Delta, Manitoba, 1938. *Trans. 4th N. Am. Wildlife Conf.*, 389–94.

———. 1994. *The canvasback on a prairie marsh*. Am. Wildlife Inst., Washington, D.C.

Hohman, W. L., C. D. Ankney, and D. L. Roster. 1992. Body condition, food habits and molt status of late-wintering ruddy ducks in California. *Southwestern Nat.* 37:268–73.

Höhn, E. O. 1975. Notes on black-headed ducks, painted snipe, and spotted tinamous. *Auk* 92:566–75.

Hoppe, R. T., L. M. Smith, and D. B. Webster. 1986. Foods of wintering diving ducks in South Carolina. *J. Field Ornith.* 57:126–34.

Hudson, R. 1976. Ruddy ducks in Britain. *Brit. Birds* 69:132–43.

Hughes, B. 1990. The ecology and behaviour of the North American ruddy duck *Oxyura jamaicensis* in Great Britain and its interactions with native waterbirds: A progress report. *Wildfowl* 41:133–38.

———. 1991. The status of the North American ruddy duck in Great Britain.

In *Britain's birds in 1989/90: The conservation and monitoring review.*, ed. D. A. Stroud and D. Glue, 162–93. BTO/NCC, Thetford.

———. 1992. The ecology and behaviour of the ruddy duck *Oxyura jamaicensis jamaicensis* (Gmelin) in Great Britain. Ph.D. diss., Univ. Bristol, England.

———, and M. Grussu. The ruddy duck in Europe and the threat to the white-headed duck. Unpublished ms.

Humphrey, P. S., D. Bridge, P. W. Reynolds, and R. T. Peterson. 1970. *Birds of Isla Grande, Tierra del Fuego.* Smithsonian Inst., Washington, D.C.

———, and G. A. K. Clark. 1964. The anatomy of waterfowl. *The waterfowl of the world,* ed. J. Delacour, Vol. 4, 167–232. Country Life, London.

———, and B. C. Livezey. 1982. Flightlessness in flying steamer ducks. *Auk* 99:368–72.

———, and K. C. Parkes. 1959. An approach to the study of molts and plumages. *Auk* 76:1–31.

Jahn, L. R., and R. A. Hunt. 1964. Duck and coot ecology and management in Wisconsin. *Wisc. Cons. Dept. Tech. Bull.* 33:1–212.

Jenni, D. 1969. Diving times of the least grebe and masked duck. *Auk* 86: 355–56.

Johnsgard, P. A. 1955. Courtship activities of the Anatidae in eastern Washington. *Condor* 57:17–27.

———. 1960a. Comparative behaviour of the Anatidae and its evolutionary implications. *Wildfowl Trust Ann. Rept.* 11:31–45.

———. 1960b. Hybridization in the Anatidae and its taxonomic implications. *Condor* 62:25–33.

———. 1961a. The taxonomy of the Anatidae—A behavioural analysis. *Ibis* 103:71–85.

———. 1961b. Tracheal anatomy of the Anatidae and its taxonomic significance. *Wildfowl Trust Ann. Rept.* 12:58–69.

———. 1962. Evolutionary trends in the behaviour and morphology of the Anatidae. *Wildfowl Trust Ann. Rept.* 13:130–48.

———. 1963. Behavioral isolating mechanisms in the family Anatidae. *Proc. XIIIth Int. Ornith. Congress, Ithaca, N.Y.,* 531–43.

———. 1964. Comparative behavior and relationships of the eiders. *Condor* 66:113–29.

———. 1965a. *Handbook of waterfowl behavior.* Cornell Univ. Press, Ithaca.

———. 1965b. Observations on some aberrant Australian Anatidae. *Wildfowl Trust Ann. Rept.* 16:73–83.

———. 1966. Behavior of the Australian musk duck and blue-billed duck. *Auk* 83:98–110.

———. 1967. Observations on the behavior and relationships of the white-backed duck and the stiff–tailed ducks. *Wildfowl Trust Ann. Rept.* 18: 98–107.

———. 1968a. Some observations on maccoa duck behavior. *Ostrich* 39:219–22.

———. 1968b. *Waterfowl: Their biology and natural history.* Univ. Nebr. Press, Lincoln.

———. 1972. *Grouse and quails of North America.* Univ. Nebr. Press, Lincoln.

———. 1973. Proximate and ultimate determinants of clutch size in the Anatidae. *Wildfowl* 24:144–49.

———. 1975. *Waterfowl of North America.* Univ. Nebr. Press, Lincoln.

———. 1978. *Ducks, geese and swans of the world.* Univ. Nebr. Press, Lincoln.

———. 1979. Order Anseriformes. In *Check-list of birds of the world,* Vol. 1, 2d. ed., ed. E. Mayr and C. W. Cottrell, 425–506. Mus. of Comp. Zoology, Cambridge, Mass.

———. 1992. *Ducks in the wild: Conserving waterfowl and their wetland habitats.* Prentice-Hall, Englewood Cliffs, N.J.

————. 1994. *Arena birds: Sexual selection and behavior.* Smithsonian Inst. Press, Washington, D.C. (in press).

————, and I. O. Buss. 1956. Waterfowl sex ratios during spring in Washington State and their interpretation. *J. Wildlife Mgmt.* 20:384–88.

————, and D. Hagemeyer. 1969. The masked duck in the United States. *Auk* 84:691–95.

————, and C. Nordeen. 1981. Display behavior and relationships of the Argentine blue-billed duck. *Wildfowl* 32:5–11.

Johnson, A. W. 1965. *The birds of Chile.* Vol. I Platt Establecimientos Graficos, Buenos Aires.

Joyner, D. E. 1969. A survey of the ecology and behavior of the ruddy duck (*Oxyura jamaicensis*) in northern Utah. M.S. thesis, Univ. Utah, Salt Lake City.

————. 1973. Interspecific nest parasitism by ducks and coots in Utah. *Auk* 90:692–93.

————. 1975. Nest parasitism and brood-related behavior of the ruddy duck (*Oxyura jamaicensis*). Ph.D. diss., Univ. Nebr., Lincoln.

————. 1977a. Behavior of ruddy duck broods in Utah. *Auk* 94:343–49.

————. 1977b. Nest desertion by ruddy ducks in Utah. *Bird-Banding* 48: 19–24.

————. 1983. Parasitic egg laying in redheads and ruddy ducks in Utah: Incidence and success. *Auk* 100:717–25.

Kear, J. 1967. Notes on the eggs and downy young of *Thalassornis leuconotus*. *Ostrich* 38:227–29.

————. 1970. The adaptive radiation of parental care in waterfowl. In *Social behaviour in birds and mammals,* ed. J. H. Crook, 358–92. Academic Press, N.Y.

Keith, L. B. 1961. A study of waterfowl ecology on small impoundments in southeastern Manitoba. *Wildlife Monogr.* 6:1–88.

King, B. 1976. Association between male North American ruddy duck and stray ducklings. *Brit. Birds* 69:34.

King, W. B. 1990. *Endangered birds of the world: the ICBP bird red data book.* Smithsonian Inst. Press, Washington, D.C.

Kortright, F. 1943. *The ducks, geese and swans of North America.* Wildlife Mgmt. Inst., Washington, D.C.

Krapu, G. L., and K. J. Reinecke. 1992. Foraging ecology and nutrition. In *Ecology and management of breeding waterfowl,* ed. B. D. J. Batt et al., 1–29. Univ. Minn. Press, Minneapolis.

Lack, D. 1967. The significance of clutch-size in waterfowl. *Wildfowl* 18:125–28.

————. 1968a. *Ecological adaptations for breeding in birds.* Methuen & Co. Ltd., London.

————. 1968b. The proportion of yolk in the eggs of waterfowl. *Wildfowl* 19: 67–69.

————. 1974a. *Evolution illustrated by waterfowl.* Blackwell Scientific Publ., Oxford.

————. 1974b. The significance of clutch size in waterfowl. *Wildfowl Trust Ann. Rept.* 18:125–28.

Ladhams, D. E. 1977. Behaviour of the ruddy duck in Avon. *Brit. Birds* 70: 137–46.

Lees-May, N. 1974. Egg of maccoa duck in the nest of the red-knobbed coot. *Ostrich* 45:39–40.

Lehmann, F. C. 1946. Two new birds from the Andes of Colombia. *Auk* 63:218–23.

Leopold, A. S. 1959. *Wildlife of Mexico: The game birds and mammals.* Univ. Calif. Press, Berkeley.

Libby, H. J. 1972. Ruddy duck distribution in relation to marsh habitat. M.S. thesis, Univ. Wisc., Madison.

Livezey, B. C. 1986. A phylogenetic analysis of Recent anseriform genera using morphological characters. *Auk* 103:737–54.

———. 1995. Phylogeny and comparative ecology of stiff-tailed ducks (Anatidae: Oxyurini). *Wilson Bull.* 107:214–34.

Lorenz, K. 1941. Comparative studies on the behaviour of the Anatinae. *Avicult. Mag.* 57:157–82; 58:8–17, 61–72, 86–94, 172–84; 59:24–34, 80–91.

Low, J. P. 1941. Nesting of the ruddy duck in Iowa. *Auk* 58:506–17.

Lowe, V. T. 1966. Notes on the musk duck, *Biziura lobata. Emu* 65:279–90.

McKinney, F. 1965. The comfort movements of Anatidae. *Behaviour* 25: 120–220.

———. 1992. Courtship, pair formation and signal systems. In *Ecology and management of breeding waterfowl,* ed. B. D. J. Batt et al., 214–50. Univ. Minn. Press, Minneapolis.

McNeil, R., P. Drapeau, and J. D. Goss–Custard. 1992. The occurrence and adaptive significance of nocturnal habits in waterfowl. *Biol. Rev.* 67: 381–419.

McKnight, D. E. 1974. Dry-land nesting by redheads and ruddy ducks. *J. Wildlife Mgmt.* 38:112–19.

Macnae, W. 1959. Notes on the biology of maccoa duck. *Bokmakierie* 11: 49–52.

Madge, S. 1984. White-headed duck with black head. *Brit. Birds* 77:154.

———, and H. Burn. 1988. *Wildfowl: An identification guide to the ducks, geese and swans of the world.* Christopher Helm, London.

Madsen, C. S., K. P. McHugh, and S. R. de Kloet. 1988. A partial classification of waterfowl (Anatidae) based on single-copy DNA. *Auk* 105:452–59.

Marchant, S., and P. Higgins (eds.). 1990. *Handbook of Australian, New Zealand, and Antarctic birds.* Vol. 1 (Ratites to Ducks). Oxford Univ. Press, Melbourne.

Matthews, G. V. T., and M. E. Evans. 1974. On the behaviour of the white-headed duck with special reference to breeding. *Wildfowl* 25:55–66.

Mendez, E. 1979. *Las aves de caza de Panama.* Privately published, Panama City.

Miller, A., and B. Collins. 1954. A nesting study of ducks and coots on Tule Lake and Lower Klamath National Wildlife Refuges. *Calif. Fish & Game* 40:17–37.

Miller, M. K., R. M. McLandress, and B. J. Gray. 1977. The display flight of the North American ruddy duck. *Auk* 94:140–42.

Misterek, D. 1974. The breeding ecology of the ruddy duck *(Oxyura jamaicensis)* on Rush Lake, Winnebago County, Wisconsin. M.S. thesis, Univ. Wisc., Oshkosh.

Morris, D. 1957. "Typical intensity" and its relation to the problem of ritualisation. *Behaviour* 11:1–12.

Murton, R. K. and J. Kear. 1976. The role of daylength in regulating the breeding seasons and distribution of wildfowl. *Symp. Brit. Ecol. Soc.* 16:337–60.

———, and ———. 1978. Photoperiodism in waterfowl: Phasing of breeding cycles and zoogeography. *J. Zool.* (London) 186:243–83.

Nores, M., and D. Yzurieta. 1980. *Aves de ambientes acuaticos de Cordoba y centro de Argentina.* Secr. de Estado de Agric. y Ganad., Cordoba.

Norman, F., and F. C. Norris. 1982. Some notes on freckled duck shot in Victoria, Australia, 1981. *Wildfowl* 23:81–7.

Oberholser, H. C. 1974. *The bird life of Texas.* Vol. I. Univ. Texas Press, Austin.

Ogilvie, M. A. 1975a. *Ducks of Britain and Europe.* T. and A. D. Poyser, Berkhamsted, U.K.

———. 1975b. The musk duck. *Wildfowl* 26:113.

Oring, L. W., and R. D. Salyer. 1992. The mating systems of waterfowl. In *Ecology and management of breeding waterfowl,* ed. B. D. J. Batt et al., 190–213. Univ. Minn. Press, Minneapolis.

Owen, M. 1977. *Wildfowl of Europe.* Macmillan, London.

———, G. L. Atkinson-Willes, and D. G. Salmon. 1986. *Wildfowl in Great Britain.* 2d ed. Cambridge Univ. Press, Cambridge.

———, and J. M. Black. 1990. *Waterfowl ecology.* Chapman & Hall, N.Y.

Palmer, R. S. (ed.). 1976. *Handbook of North American birds.* Vol. 3. Yale Univ. Press, New Haven, Conn.

Patton, J. C., and J. C. Avise. 1983. An empirical evaluation of qualitative Hennigian analysis of protein electrophoretic data. *J. Mol. Biol.* 19:244–54.

Peck, C. K., and R. D. James. 1983. *Breeding birds of Ontario: Nidiology and distribution.* Vol. 1: *Non-passerines.* Royal Ontario Mus., Toronto.

Peña, M. R. de la. 1976. *Aves de la provincia de Santa Fé, Castellví, S. A.* Sante Fé.

Peters, J. 1931. *A check-list of the birds of the world.* Vol. 1. Harvard Univ. Press, Cambridge.

Phillips, J. C. 1922–26. *A natural history of the ducks.* 4 vols. Houghton Mifflin, Boston. (Reprinted as 2 vols. in 1986 by Dover Books, N.Y.)

Poole, E. L. 1938. Weights and wing areas in North American birds. *Auk* 55:511–17.

Powell, A. 1979. Cuckoo in the nest. *Wildfowl News* 81:15–16.

Pyecraft, W. P. 1906. Notes on a skeleton of the musk duck, *Biziura lobata,* with special reference to skeletal characteristics evolved in relation to the diving habits of the bird. *J. Linn. Soc.* (London) *Zool.* 29:396–407.

Raffaele, H. 1983. *A guide to the birds of Puerto Rico and the Virgin Islands.* Fundo Educatavo Interamericano, San Juan.

Raikow, R. J. 1970. Evolution of diving adaptations in the stifftail ducks. *Univ. Calif. Publ. Zool.* 94:1–52.

———. 1971. The osteology and taxonomic position of the white-backed duck, *Thalassornis leuconotus. Wilson Bull.* 83:270–77.

Rees, E. C., and N. Hillgarth. 1984. The breeding biology of the black-headed duck and the behavior of their young. *Condor* 86:242–50.

Ridgway, R. 1880. Revision of nomenclature of North American birds. *Proc. U.S. Natl. Mus.* 3:15–16.

Rieneker, N. C., and W. Anderson. 1960. A waterfowl nesting study on Tule Lake and Lower Klamath National Wildlife Refuges, 1957. *Calif. Fish & Game* 46:481–506.

Ripley, S. D., and G. E. Watson. 1956. Cuban bird notes. *Postilla* 26:1–6.

Roberts, T. S. 1932. *The birds of Minnesota.* Vol. 1. Univ. Minn. Press, Minneapolis.

Robinson, F. N., and A. H. Robinson. 1970. Regional variation in the visual and acoustic signals of the male musk duck, *Biziura lobata. CSIRO Wildlife Res.* 15:73–78.

Rogers, M. 1977. Ruddy ducks on tidal waters. *Brit. Birds* 70:219–20.

Rohwer, F. C. 1988. Inter- and intraspecific relationships between egg size and clutch size in waterfowl. *Auk* 105:161–76.

———. 1992. The evolution of reproductive patterns in waterfowl. In *Ecology and management of breeding waterfowl,* ed. D. J. Batt et al., 486–539, Univ. Minn. Press, Minneapolis.

———, and S. Freeman. 1989. The distribution of conspecific nest parasitism in birds. *Can. J. Zool.* 67:239–53.

Root, T. 1988. *Atlas of wintering North American birds: An analysis of Christmas Bird Count data.* Univ. Chicago Press, Chicago.

Salvadori, T. 1895. Catalogue of the Chenomorphae (Palamedeae, Phoenicopteri, Anseres), Crypturi, and Ratitae in the collection of the British Museum. *Catalogue of the birds of the British Museum* 27:1–636.

Salyer, R. D. 1992. Ecology and evolution of brood parasitism in waterfowl. In *Ecology and management of breeding waterfowl*, ed. (B. D. J. Batt et al., 290–323. Univ. Minn. Press, Minneapolis.

Savage, C. 1965. White-headed ducks in Pakistan. *Wildfowl Trust Ann. Rept.* 16:121–26.

Sclater, P. L. 1880. List of the certainly known species of Anatidae, with notes on such as have been introduced into the zoological gardens of Europe, and remarks on their distribution. *Proc. Zool. Soc. London*, 1880:496–536.

Scott, D., and M. Carbonell. 1986. *Inventario de humedales de la region Neotropical.* International Wildfowl Research Bureau (IWRB), Slimbridge, and International Union for the Conservation of Nature (IUCN), Cambridge, U.K.

Scott, P. 1958. Notes on Anatidae seen on world tour. *Wildfowl Trust Ann. Rept.* 9:69–72.

Seaman, G. A. 1958. Masked duck collected in St. Croix, Virgin Islands. *Auk* 75:215.

Serventy, D. L. 1946. Display of the musk duck in western Australia. *Emu* 45:318–21.

Shanks, D. 1954. A further note on blue-billed ducks. *Emu* 54:76.

Short, L. L. 1976. Notes on a collection of birds from the Parguayan chaco. *Amer. Mus. Novitates* 2597:1–16.

Shufeldt, R. W. 1914. Contribution to the study of the "tree-ducks" of the genus *Dendrocygna. Zool. Jahrb. (Syst., Geog. & Biol.)* 38:1–70.

Sibley, C. G., and J. E. Alquist. 1990. *Phylogeny and classification of birds: A study in molecular evolution.* Yale Univ. Press, New Haven, Conn.

———, and B. L. Monroe, Jr. 1990. *Distribution and taxonomy of birds of the world.* Yale Univ. Press, New Haven, Conn.

Siegfried, W. R. 1964. Parasitic egg-laying in South African Anatidae. *Ostrich* 35:61.

———. 1968. Non-breeding plumage in the adult maccoa duck. *Ostrich* 39:91–93.

———. 1969a. Breeding season of the maccoa duck in the south western Cape. *Ostrich* 40:213.

———. 1969b. The proportion of yolk in the egg of the maccoa duck. *Wildfowl* 20:78.

———. 1970. Double wing moult in the maccoa duck. *Wildfowl* 21:122.

———. 1973a. Platform building by male and female ruddy ducks. *Wildfowl* 24:150–53.

———. 1973b. Post-embryonic development of the ruddy duck *Oxyura jamaicensis* and some other diving ducks. *Intern. Zoo Yearbook* 13:77–87.

———. 1973c. Summer foods and feeding of the ruddy duck in Manitoba. *Canad. J. Zool.* 51:1293–97.

———. 1973d. Wing moult of ruddy ducks in Manitoba. *Bull. Brit. Orn. Club* 93:98–99.

———. 1976a. Breeding biology and parasitism in the ruddy duck. *Wilson Bull.* 88:566–74.

———. 1976b. Segregation of feeding behavior in four diving ducks. *Can. J. Zool.* 54:730–36.

———. 1976c. Social organization in ruddy and maccoa ducks. *Auk* 93:560–70.

———. 1977. Notes on the behaviour of ruddy ducks during the brood period. *Wildfowl* 29:126–28.

———. 1985. Socially induced suppression of breeding plumage in the maccoa duck. *Wildfowl* 36:135–37.

————, A. E. Burger, and P. J. Caldwell. 1976. Incubation behavior of ruddy and maccoa ducks. *Condor* 78:512–17.

————, ————, and P. G. H. Frost. 1976. Energy requirements for breeding in the maccoa duck. *Ardea* 64:171–91.

————, ————, and F. J. van der Merwe. 1976. Diurnal time budget of territorial male maccoa ducks, *Oxyura maccoa*. *Zool. Afr.* 11:111–25.

————, and F. van der Merwe. 1975. A description and inventory of the displays of the maccoa duck. *Zeitsch. f. Tierpsychol.* 37:1–23.

Smart, G. 1965. Body weights of newly hatched Anatidae. *Auk* 82:645–48.

Smith, A. G. 1949. Migration of the ruddy duck. U.S. Fish & Wildlife Serv. *Spec. Sci. Rept.: Wildlife* 1:45–46.

Smith, L. M., L. D. Vanglider, R. T. Hoppe, S. J. Morreale, and I. L. Brisbin, Jr. 1986. Effect of diving ducks on benthic food resources during winter in South Carolina, U.S.A. *Wildfowl* 37:136–41.

Snyder, D. E. 1966. *The birds of Guyana*. Peabody Museum, Salem, Mass.

Stark, R. T. 1978. Food habits of the ruddy duck *(Oxyura jamaicensis)* at the Tinicum National Environmental Center. M.S. thesis, Penn. State Univ., State College.

Steel, P. E., P. D. Dalke, and E. G. Bizeau. 1956. Duck production at Gray's Lake, Idaho. *J. Wildlife Mgmt.* 20:279–85.

Stewart, R. 1962. Waterfowl populations in the upper Chesapeake region. U.S. Fish & Wildlife Serv. *Spec. Sci. Rept.: Wildlife* 65:1–208.

Stranger, R. H. 1961. Display of the musk duck. *W. Australian Nat.* 7:210–11.

Sylvestre, S. de V. 1993. The spread of the ruddy duck in Spain and its impact on the white-headed duck. *IWRB Threatened Waterfowl Research Group Newsletter.* 3:3–4.

Tinbergen, N. 1954. The origin and evolution of courtship and threat display. *In Evolution as a process*, ed. J. Huxley, A. C. Hardy and E. B. Ford, 233–50. Geo. Allen & Unwin Ltd., London.

Todd, F. S. 1979. *Waterfowl: Ducks, geese and swans of the world*. Sea World Press, San Diego.

Tome, M. E. 1981. Reproductive bioenergetics of female ruddy ducks in Manitoba. M.S. thesis, Univ. Maine, Orono.

————. 1984. Changes in the nutrient reserves and organ size of female ruddy ducks in Manitoba. *Auk* 101:830–37.

————. 1987. An observation of renesting by a ruddy duck, *Oxyura jamaicensis*. *Can. Field-Nat.* 101:153–54.

————. 1988. Optimal foraging: Food patch depletion by ruddy ducks. *Oecologia* 76:27–36.

————. 1989. Food patch depletion by ruddy ducks: Foraging by expectation rules. *Can. J. Zool.* 67:2751–55.

————. 1991. Diurnal activity budget of female ruddy ducks breeding in Manitoba. *Wilson Bull.* 103:183–89.

————, and D. A. Wrubleski. 1988. Underwater foraging behavior of canvasbacks, lesser scaups and ruddy ducks. *Condor* 90:168–72.

Torres, J. A., and R. Arenas. 1983. Characteristicas de la poblacion Española de malvasias *(Oxyura leucocephala)* durant el año 1983. *Oxyura* 1:5–19. (English summary).

————, and ————. 1985. Nuevos datos relativos a la alimentacion de *Oxyura leucocephala*. *Ardeola* 32:127–131. (English summary.)

————, ————, and J. M. Ayala. 1986. Evolution historica de poblacion Española de malvasia *(Oxyura leucocephala)*. *Oxyura* 3:5–17.

————, and J. M. Ayala. 1986. Variation in head color of the white-headed duck. *Alauda* 34:197–206. (In French, English summary.)

————, and C. Raya Gomez. 1982. La reproduccjon de la malvasia *(Oxyura leucocephala)* en el sur de la provincia de Cordobá, España. Donaña, *Acta. Vert.* 10:122–31. (English summary).

——— and ——— 1983. La reproducción de la malvasía *(Oxyura leucocephala)* en el sur de la provincia de Cordoba. España. *Donafia. Acta Vert.* 10:123–31.

———, ———, R. M. Arenas, and J. M. Ayala. 1985. Estudio del comportamiento reproductor de la malvasía *(Oxyura leucocephala). Oxyura* 2:5–21. (English summary).

de Vida, S. 1993. The spread of the ruddy duck in Spain and its impact on the white-headed duck. *IWRB Threatened Waterfowl Research Group Newsletter* 3:3–4.

Veselovsky, Z. 1976. Beitrage zur Kenntnis der Ruderente, *Oxyura leucocephala* (Scopoli 1769). *Beitr. Vogelk.* (Leipzig) 22:105–14.

Weller, M. W. 1959. Parasitic egg laying in the readhead *(Aythya americana)* and other North American Anatidae. *Ecol. Monogr.* 29:333–65.

———. 1967a. Distribution and habitat selection of the black-headed duck. *Hornero* 10:299–306.

———. 1967b. Notes on some marsh birds of Cape San Antonio, Argentina. *Ibis* 109:391–411.

———. 1967c. Notes on the plumages and weights of the black-headed duck, *Heteronetta atricapilla. Condor* 69:133–45.

———. 1968a. The breeding biology of the parasitic black-headed duck. *Living Bird* 7:169–207.

———. 1968b. Notes on some Argentine anatids. *Wilson Bull.* 80:189–212.

Wetmore, A. 1917. On certain secondary sexual characteristics in the male ruddy duck, *Erismatura jamaicensis. U.S. Nat. Mus. Proc.* 219:479–82.

———. 1918. A note on the tracheal air-sac of the ruddy duck. *Condor* 20:19–20.

———. 1926. Observations on the birds of Argentina, Paraguay, Uruguay and Chile. *U. S. Nat. Mus. Bull.* 133:1–488.

———. 1965. *The birds of the Republic of Panama.* Pt. 1. *Tinamidae (tinamous) to Rhynchopidae (skimmers). Smithsonian Misc. Coll.* 150:1–483.

Wheeler, J. R. 1953. Notes on the blue-billed ducks at Lake Wendouree, Ballarat. *Emu* 53:280–82.

Williams, C. S., and W. H. Marshall. 1938. Duck nesting studies, Bear River Migratory Bird Refuge, Utah, 1937. *J. Wildlife Mgmt.,* 2:29–48.

Wintle, C. C. 1981. Notes on the breeding behaviour of the white-backed duck. *Honeyguide* 105:13–20.

Woodin, M. C. 1987. Wetland selection and foraging ecology of breeding diving ducks. Ph.D. diss., Univ. Minn., St. Paul.

———, and G. A. Swanson. 1989. Foods and dietary strategies of parasitic-nesting ruddy ducks and redheads. *Condor* 91:280–87.

Woolfenden, G. 1961. Postcranial osteology of the waterfowl. *Florida State Mus. Bull.* 6:1–129.

Ydenberg, R. C. 1986. Foraging by diving birds. *Proc. Intern. Ornith. Congr.* 19:1832–42.

Index

Note: Complete indexing for each stifftail species may be found following the species' preferred English vernacular name, as adopted in this book. The primary descriptive species account for each stifftail species is indicated by italics. Only the first author of multiple-author publications has been indexed, and most organisms other than birds are not indexed unless they are of special biological significance to stifftails.